THE AESTHETIC DIMENSION
OF SCIENCE

THE AESTHETIC
DIMENSION
OF SCIENCE

1980 NOBEL CONFERENCE

organized by

Gustavus Adolphus College
St. Peter, Minnesota

edited by

Deane W. Curtin
Department of Philosophy
Gustavus Adolphus College

Philosophical Library
New York

Library of Congress Cataloging in Publication Data

Nobel Conference (1980 : Gustavus Adolphus
  College)
  The aesthetic dimension of science.

  1. Science—Philosophy—Congresses.   2. Aesthetics
—Congresses.   I. Curtin, Deane W.   II. Gustavus
Adolphus College.   III. Title.
Q174.N6   1980   501   81-19243
ISBN 0-8022-2393-1   AACR2

Overseas distributor: George Prior Associated Publishers Ltd.
Lower Ground Floor, High Holborn House, 52-54 High Holborn,
London WC1V 6RL England

To
Thomas Gover

Contents

The Contributors

WILLIAM NUNN LIPSCOMB, Jr. is Abbott and James Lawrence Professor of Chemistry at Harvard University. He received his education at the University of Kentucky and at the California Institute of Technology where he studied with Linus Pauling.

Among his numerous honors and awards are the Nobel Prize in Chemistry (1976) for his work on boron hydrides, the Senior U.S. Scientist Award, Alexander von Humboldt-Stiftung (1979), and the Peter Debye Award in Physical Chemistry, American Chemical Society (1973).

CHEN NING YANG is Albert Einstein Professor of Physics and Director of the Institute of Theoretical Physics at the State University of New York at Stony Brook. Born in Anwhei, China, he attended National Southwest University in Kunming, China, Tsinghua University, and the University of Chicago, where he received his Ph.D. in 1948. His previous appointments include The Institute for Advanced Studies, Princeton, Gibbs Lecturer of the American Mathematical Society, and Loeb Lecturer at Harvard University.

Professor Yang was awarded the Nobel Prize in 1957 for

his work on strong and weak interactions. He was also the recipient of the Albert Einstein Commemorative Award (1957), and the Rumford Prize (1980).

FREEMAN J. DYSON is Professor of Physics, Institute for Advanced Studies, Princeton. Born in Crowthorne, Berkshire, England, he attended Winchester College and the University of Cambridge. Prior to his appointment at the Institute for Advanced Study, Professor Dyson was Professor of Physics at Cornell University, Visiting Professor at Yeshiva University, and Visiting Professor at the Max Planck Institute for Physics and Astrophysics.

Author of *Disturbing the Universe* (1979), Professor Dyson has been awarded the Lorentz Medal (1966), the J. Robert Oppenheimer Memorial Prize (1970), and the Harvey Prize (1977).

GUNTHER SCHULLER was born in New York City and educated at St. Thomas Choir School and the Manhattan School of Music. Perhaps best known as the founder and president of the New England Conservatory of Music, he has also taught at the Manhattan School of Music, and was head of the composition department at Tanglewood. In addition to his activities as composer and teacher, Mr. Schuller has conducted many of the world's great orchestras including those of Boston, Cleveland, Philadelphia, New York, Chicago, Minnesota, and the Berlin Philharmonic.

Mr. Schuller is the author of two books: *Horn Technique* (1962), and *Jazz: Its Roots and Musical Development* (1968).

CHARLES HARTSHORNE was born at Kittanning, Pennsylvania and educated at Haverford College, Harvard University, the University of Freiburg and the University of Marburg. Professor Emeritus of Philosophy at the University of Texas at Austin, Professor Hartshorne's previous appointments include Harvard University, The University of Chicago, and Emory University.

Professor Hartshorne is a prolific author whose books include *Man's Vision of God* (1941), *The Logic of Perfection* (1962), *Anselm's Discovery* (1965), and *A Natural Theology for Our Time* (1967). In addition to his work in philosophy, he is a noted ornithologist and the author of *Born to Sing: An Interpretation and World Survey of Bird Song* (1973).

Acknowledgments

The 1980 Nobel Conference was made possible by Russell T. Lund and the late Rhoda Berge Lund who made a gift in 1978 to endow the Nobel Conferences and Nobel related activities of Gustavus Adolphus College.

The editor also wishes to thank Vice President Robert Peterson and the Development Office staff as well as Professor Luke Reinsma, Professor Omer Prewitt, and Julie Schendel for their valuable assistance.

Preface

This volume is the outgrowth of a conference held at Gustavus Adolphus College, October 7 and 8, 1980, titled "The Aesthetic Dimension of Science." It was the sixteenth such conference sponsored by the college and associated with the Nobel Foundation dating back to 1965. The idea for the series was born in 1963 when, after ceremonies naming Gustavus's new science facilities in honor of Alfred Nobel, Dr. Glenn Seaborg (Nobel Prize, Chemistry, 1951) suggested to then president Dr. Edgar M. Carlson, that the college should sponsor a living memorial to Alfred Nobel in the form of a series of Nobel Conferences. Permission of the Nobel Foundation was granted, and to this day the Nobel Conference program at Gustavus Adolphus College remains the only formal Nobel program in the world outside of Sweden and Norway.

Under the direction of Professor Thomas Gover, the 1980 Nobel Conference explored the issue of whether there is an aesthetic dimension to science. The specific issues addressed included "Is beauty an essential element in scientific inquiry?" "Given the choice between a beautiful or elegant theory which appears not to fit the facts perfectly, and an unelegant theory which does appear to fit the facts, which should be adopted?" "How is creativity in the sciences like or unlike creativity in the arts or in philosophy?" And, "What specific qualities of a scientific theory, a philosophical theory, or a work of art contribute to its aesthetic worth?"

The conference began with the contributions of two Nobel laureates, William Lipscomb and C.N. Yang. Each developed an argument in favor of regarding science as a thoroughly aesthetic enterprise. While not suggesting it is the only aesthetic value in science, each emphasized the quality of symmetry, and each tended to point out the unity brought about by the advance of science.

While it would go too far to say that Professors Dyson, Schuller, and Hartshorne disagreed with the first two papers, they displayed distinctly different "tastes" in the aesthetics of science and art. Each agreed that symmetry is an aesthetic value, but chose to stress what might be termed the dynamic, unstable elements in nature, for instance, asymmetry rather than symmetry, and diversity instead of unity. Professor Dyson argued that science requires both unifiers, like Einstein, and diversifiers, like Rutherford. Placing himself in the latter camp, he said, "I gazed at the stars as a young boy. That's what science means to me. It's not theories about stars; it's the actual stars that count."

Gunther Schuller is widely recognized as an advocate of the "global view" of music. His paper begins with a point of contrast between science and music. Whereas we have little or no trouble specifying the sense in which science progresses, we find it difficult, if not impossible, to do the same for music. A consequence of the global view is that, ". . . there are only two kinds of music as Giacino Rossini put it a century and a half ago: 'good and bad.' There is no inherent or automatic correlation between *categories* or *types* of music and quality, or the lack of it." Professor Schuller concludes his paper, however, with a tentative suggestion that music and science may be moving in the same direction after all. As he says, "In my very amateur but nevertheless hopefully correct view of the history and development of science, I see an extraordinary parallel to what I have called a 'global view of music:' namely if the development of the universe can be described as a never-ending breaking up—'symmetry breaking' as you call it—of cells and molecules and atoms and microorganisms, then so too music has proliferated and fragmented in never-ending replication and differentiation from the first moment when primitive man uttered his first primeval sound."

Finally, in a characteristically rich paper, Charles Hartshorne argues that beauty is the mean between two sets of extremes: first, between mere unity and mere diversity, and second, between the ultracomplex and the ultrasimple. Having defined beauty, he contends that science is the "imaginative, observational, critical love of nature," and that far from being merely practical in its origin, it is "the search for the hidden beauty of the world."

A belief still widely held is that whereas scientific inquiry is coldly rational, artistic inquiry is rather heatedly emotional. If the participants in this conference are to be believed, this view is at least wildly exaggerated. The picture of scientific inquiry which emerged here is that of a deeply human endeavor, a picture which makes it more difficult that one might expect to distinguish the sciences and the arts as fundamentally different modes of human activity.

One thought-provoking attempt to distinguish the two by referring to the results of such inquiry was offered by Professor Yang during one of the discussions. Prompted by Professor Schuller's global view of music, he pointed out that while "there are many possible systems of music," that is, "The judgment of what is beautiful in music is not unique," the same is not true of science. Posing the question, "Are many systems of physics possible?" he answers in the negative. If there is some scientifically advanced culture, they must know Maxwell's equations in roughly the same form we do. This is only one of many intriguing topics addressed in the series of discussions which complete this volume.

THE AESTHETIC DIMENSION
OF SCIENCE

Aesthetic Aspects of Science

William N. Lipscomb

On Science

This past summer at a Meeting in Lindau, I took my seat at lunch beside a lady to whom I introduced myself. She replied, "How do you do, I am Mrs. Heisenberg." Immediately I asked, "Are you certain?"

She was. Indeed, she was the widow of Werner Heisenberg. He showed us that all measurements involve the measuring system (us), and that there are statistical aspects rather than certainties in every measurement. It is within these limits that we expect to choose among rival theories or descriptions by the use of experiments or observations in science.

But how does one arrive at the theories? I am prepared, I suppose, to recognize some truth in the methods of inductive and deductive reasoning which I was taught as the methods of science. However, it was already clear to me from original research that I had done previously in high school that I was more inclined to make large intuitive jumps, and then set about to test the conclusions. It was not until many years later upon my arrival in Graduate School at Caltech that I

1

observed Linus Pauling using much the same methods. For example, when Pauling was describing his theory of electroneutrality in molecules and reactions a student asked him to derive it. Pauling said that there was no derivation. The student then asked how he arrived at the principle. Pauling said, "I made it up!" He often refers to his methods as stochastic.

There must be some fundamental survival advantage, not completely described to us by the psychologist or anthropologist, in observation, perception, reasoning and abstraction.[1] Were myths and the arts necessary for cohesive social behavior in the evolution of man? I do not pretend to answer questions so far from my field. I do want to observe, however, that the absolutely amazing degree of abstraction in the mathematics of the last two centuries is equalled only by much later discoveries that some of these developments are relevant to the physical sciences. They form the basis for descriptions of nature at a level that could hardly have been conceived by the mathematicians themselves. For example, the studies of Riemann clarified the notions of space and gave a general theory of an unlimited number of spaces. Not only were we freed of Euclidean space as a necessary description, but Einstein showed that we really live in a particular kind of four dimensional space in which the physical laws have no fixed frame of reference. Another example is the theory of mathematical groups (Cauchy, Galois) which generalized symmetry far beyond the obvious, and which was to become part of a description of nature at a profound level.

What guides the mind in these discoveries? One might observe that Riemann and Cauchy were also physicists, although their mathematical discoveries were to influence physics of a much later day. (Galois hardly had a chance; [2] he was killed in a senseless duel at the age of 21). One knows that they were influenced by the mathematics of the time (Riemann by Gauss, for example), but it is more to the point that they were driven by their discoveries to see them through —a demon which possesses every truly original person. The developments from Euclid, Lobachevsky, Cauchy, Galois and Lie become more abstract, and then more relevant than ever to the symmetry aspects of nature.

And so, to illustrate the nature of science, I have chosen mathematicians![3] I have done so for two reasons. First, much of the abstraction that occurred in the science of this century, and which is sometimes correlated with similar changes in visual art, music, literature and other arts, actually began much earlier in mathematics. Second, the nature of theoretical science became different after Einstein and the developers of quantum mechanics (Planck, Bohr, Heisenberg, Schroedinger, Born). Physics had moved from relationships among observations, to the invariant laws of Newton, to Maxwell's second order differential equations, and finally to new kinds of invariances based partly upon symmetry arguments. While experimental or observational criteria are used to decide among rival theories, the methods for arriving at these theories of nature have much of the simplicity and beauty of that part of pure mathematics which is selected or developed for these descriptions of nature.

On Aesthetics

From my own experience, I would certainly not separate aesthetics from science. When, after years of research I realized that a whole area of chemistry (of boron) was really quite different from anything that had previously been thought, I felt a focusing of intellect and emotions which was surely an aesthetic response. It was followed by a flood of new predictions coming from my mind as if I were a bystander watching it happen.[4] Only later was I able to begin to formulate a systematic theory of structure, bonding, and reactions of these un-

3

usual molecules. Both the structures and wave functions describing the bonding were based on simple polyhedra of high symmetry and their fragments. Was it science? Our later tests showed that it was. But the processes that I used and the responses that I felt were more like those of an artist.

Aesthetics, certainly among the most complex responses of the human mind, may be easier to describe than to define. If what follows does not agree with your view of aesthetics, I can only appeal to Lewis Carroll: [5] "When I use a word," Humpty Dumpty said, in a rather scornful tone, "it means just what I choose it to mean—neither more nor less."

The wider, more recent view, of the discipline of aesthetics includes its original meaning as a systematic exploration of beauty in the arts. Beauty will be taken here as the proper conformity of the parts of a work of art to one another, and to the whole.[6] Aesthetics goes beyond the creation or enjoyment of a particular work of art to more general responses, explanations and evaluations of merit. The conceptual aspects include aesthetic appreciation of the arts as they are created (or selected) by man; these concepts are sometimes organized after the model that some have of the sciences. In addition, a recognition of the essence and vivid quality of the work of art or event must be present, as distinct from the appearance or the particular quality. The artist must have assimilated, transformed and integrated an experience, event expression, feeling or deep emotion, and then presented the result for response in others.[7] Those who respond should value the experience or object for its own sake, and they can do so either conceptually or intuitively.

Having stated this view of aesthetic criteria, I would consider it too ambitious to draw extended comparisons between aesthetics in the arts and in the sciences. I would rather try to convey to you an appreciation of some aesthetic aspects of science which seem especially clear to me. I begin with quotations relating to beauty, creativity, simplicity and aesthetics,

4

and then turn to the development of ideas of symmetry and asymmetry, in nature.

Quotations on Beauty and Aesthetics in Science

In his Nora and Edward Ryerson Lecture of 1975, Chandrasekhar [8] gives some illuminating examples of beauty in science.
Keplar:

"Mathematics is the archetype of the beautiful."

David Hilbert (in his memorial address for Hermann Minkowski):

"Our Science, which we loved above everything, had brought us together. It appeared to us as a flowering garden. In this garden there were wellworn paths where one might look around at leisure and enjoy oneself without effort, especially at the side of a congenial companion. But we also liked to seek out hidden trails and discovered many an unexpected view which was pleasing to our eyes; and when the one pointed it out to the other, and we admired it together, our joy was complete."

Hermann Weyl:

"My work always tried to unite the true with the beautiful; but when I had to choose one or the other, I usually chose the beautiful." [9]

Heisenberg (in a discussion with Einstein):

"If nature leads us to mathematical forms of great simplicity and beauty—by forms I am referring to coherent

5

systems of hypothesis, axioms, etc.—to forms that no one has previously encountered, we cannot help thinking that they are 'true,' that they reveal a genuine feature of nature. . . . You must have felt this too: the almost frightening simplicity and wholeness of the relationships which nature suddenly spreads out before us and for which none of us was in the least prepared."

Towards the end of May 1925, Heisenberg, ill with hay fever, went to Heligoland to be away from flowers and fields. There by the sea, he made rapid progress in resolving the difficulties in the quantum theory as it was at that time. He writes:

"Within a few days more, it had become clear to me what precisely had to take the place of the Bohr-Sommerfeld quantum conditions in an atomic physics working with none but observable magnitudes. It also became obvious that with this additional assumption, I had introduced a crucial restriction into the theory. Then I noticed that there was no guarantee that . . . the principle of the conservation of energy would apply . . . Hence I concentrated on demonstrating that the conservation law held; and one evening I reached the point where I was ready to determine the individual terms in the energy table [Energy Matrix] . . . When the first terms seemed to accord with the energy principle, I became rather excited, and I began to make countless arithmetical errors. As a result, it was almost three o'clock in the morning before the final result of my computations lay before me. The energy principle had held for all the terms, and I could no longer doubt the mathematical consistency and coherence of the kind of quantum mechanics to which my calculations pointed. At first, I was deeply alarmed. I had the feeling that, through the surface of atomic phenomena, I was looking at a strangely beautiful interior, and felt almost giddy at the thought that I now had to probe this wealth of mathematical structure nature had so generously spread out before me. I was far too excited to sleep, and so, as a new day dawned, I made for the southern tip of the island, where I had been longing to climb a rock jutting out into the sea. I now did so without too much trouble, and waited for the sun to rise."

Chandrasekhar:

"In my entire scientific life, extending over forty-five years, the most shattering experience has been the realization that an exact solution of Einstein's equations of general relativity, discovered by the New Zealand mathematician, Roy Kerr, provides the *absolutely exact representation* [10] of untold numbers of massive black holes that populate the universe. This 'shuddering before the beautiful,' this incredible fact that a discovery motivated by a search after the beautiful in mathematics should find its exact replica in Nature, persuades me to say that beauty is that to which the human mind responds at its deepest and most profound."

Poincaré:

"The Scientist does not study nature because it is useful to do so. He studies it because he takes pleasure in it; and he takes pleasure in it because it is beautiful. If nature were not beautiful, it would not be worth knowing and life would not be worth living. . . . I mean the intimate beauty which comes from the harmonious order of its parts and which a pure intelligence can grasp."

Commenting on these observations of Poincaré, J.W.N. Sullivan, the author of perceptive biographies of both Newton and Beethoven wrote (in the *Athenaeum* for May 1919):

"Since the primary object of the scientific theory is to express the harmonies which are found to exist in nature, we see at once that these theories must have an aesthetic value. The measure of the success of a scientific theory is, in fact, a measure of its aesthetic value, since it is a measure of the extent to which it has introduced harmony in what was before chaos.

"It is in its aesthetic value that the justification of the scientific theory is to be found, and with it the justification of the scientific method. Since facts without laws would be of no interest, and laws without theories would have, at most, only a practical utility, we see that the motives which guide the scientific man are, from the beginning, manifestations of the aesthetic impulse. . . . The measure in which science falls short of art is the measure in which it is incomplete as science. . . ."

A most interesting example of a theory which was kept alive because of its beauty, even though it was wrong, is Weyl's gauge theory of gravitation, which resulted from an attempt to include electromagnetism.[11] Weyl suggested a geometrical (gauge) invariance for the electromagnetic potential A_μ, modelled after the coordinate invariance of the theory of gravity. However, the theory did not describe electromagnetism properly. After the invention of quantum mechanics, London in 1927 identified Weyl's scale factor with the imaginary $-ieA_\mu/\hbar c$ rather than with A_μ, and hence the gauge invariance became a local phase invariance, which correctly described electromagnetism. Weyl's instincts were absolutely remarkable.

Einstein, in reference to Bohr's work following Planck's law and his own related work on the photoelectric effect,[12] said

"That this insecure and contradictory foundation was sufficient to enable a man of Bohr's unique instinct and tact to discover the major laws of the spectral lines and of the electron shells of the atoms together with their significance to chemistry appeared to me like a miracle— and appears to me as a miracle even today. This is the highest form of musicality in the sphere of thought."

J. Larmor on scientific progress, in the preface to Poincaré's "Science and Hypothesis":

"New ideas emerge dimly into intuition, come into consciousness from nobody knows where, and become the material on which the mind operates, forming them gradually into a consistent doctrine, which can be welded onto existing domains of knowledge. But this process is never complete: a crude connection can always be pointed to by a logician as an indication of the imperfection of human constructions."

Ilse Rosenthal-Schneider (in reference 12, p. 145):

"The deep satisfaction found in scientific work, akin to the delight derived from genuine art, is one of the fundamental human emotions which is highly intensified by personal contact with the creative mind."

Perhaps these comments will illustrate that beauty and simplicity, or imagination and creativity have some things in common. In themselves, or in their attention to particular cases they do not create a discipline of aesthetics. The aesthetic response transcends particular examples, although we are made aware of this response by attention to a particular work of art or science. I would like now to turn to a part of science which may serve to illustrate an aspect of aesthetics.

Symmetry

Of the many choices of aesthetic criteria available in the arts, symmetry is only one. However, it is a very natural choice for an aesthetic aspect of science. Symmetry is more often perceived than learned, although it can be described. Sometimes it is so obvious that deviations from symmetry are required in order to maintain the aesthetic interest. And sometimes symmetry is so subtle that it can hardly be recognized. In some examples I shall use the term more generally as approximate balance in others more strictly as pure symmetry. Nor shall I neglect complete asymmetry.

Symmetry takes on such a fundamental and profound significance that the physical scientist of this century now extracts the absolute maximum that can be obtained from symmetry, often before beginning the other aspects of the problem. When symmetries are broken beyond easy recognition, their discoveries can be long delayed. And we shall see that some of the symmetries of science are very different from the usual two- and three- dimensional space symmetries of the arts.

Nevertheless, let us begin with three-dimensional space. Moving from the (apparent only) bilateral symmetry of animals, including even the starfish, we come to a problem connected with one of the great unknowns: how did life start? Specifically, why does the genetic code yield amino acids of one hand only, the L-configuration? [13] Indeed, when a spark discharge is passed through a mixture of gases thought to be in the earth's primordial (reducing) atmosphere both D- and L-amino acids have been shown to be present. [14] Special mechanisms have often been invented to explain life's pref-

10

erence for these left-handed protein molecules, probably formed by a primitive genetic code. However, it is likely that a very small imbalance could cause one hand to win over the other in the billions of years during which life has existed on the earth. Nevertheless, the observation of only L-amino acids in the genetic code is very disturbing to one's aesthetic sense of symmetry. At present we really do not know whether a form of life based upon D-amino acids existed in the earth's early history.

These L-amino acids form the basis for the structures of proteins. Both Pauling's α-helix of proteins and the double helical structure of the genetic material DNA proposed by Watson and Crick have aesthetic appeal, not only in their atomic arrangements which are elegant indeed, but also in the enormous development of science which has been spawned by these proposals. One had been guided through a high narrow pass through the mountains, and then shown suddenly a view of a new, fertile meadow. Crick has commented that both symmetry and simplicity played roles in the discovery of the double helix.[15]

A more detailed look at three dimensional symmetry may show its depth in science. Consider a symmetry operation, which leaves the object apparently unchanged. An example is the rotation of a square in its plane by 90°. A symmetry operation will superimpose the transformed object onto the original object, or onto its mirror image when this is feasible. If a point in space remains in a fixed position, the former operation is a rotation; and if a translation in space is also required, for example in a repeating structure such as a crystal, the single operation is called a screw rotation. When the object can be superimposed on its mirror image, the symmetry operation is a rotatory reflection. This single operation is a rotation followed or preceded by a reflection in a plane perpendicular to the axis of rotation.[16] (The rotatory inversion is completely equivalent. Also it is often convenient to keep the gliding reflection as a special case.) When these operations

11

are combined with repetitive translations in space, there are 230 types of symmetry. These symmetry types describe all known crystal structures. Only 1, 2, 3, 4 and 6 fold axes are found. No crystal can have a five fold axis, nor can it have a 7 or higher fold axis. In two dimensions, there are 17 ways of combining these other symmetries (axes, mirrors, glides) with repetitive translations. It is remarkable that all 17 were discovered empirically by the Arabic artists of the Alhambra in Granada.[17]

Extensions of three-dimensional symmetry are known. Antisymmetry, introduced by Heesch in 1929 and Shubnikov in 1945, includes the black and white groups where these two colors are related by antisymmetry.[18] An extension of polychromatic symmetry up to twelve colors is beautifully displayed in two dimensions by Loeb.[19] These color groups are useful in describing magnetic properties. For example, hematite (α-Fe_2O_3) between -20 and $675°$ has a structure, including the magnetic aspects, which is based upon a three-color group. Also, in the analysis of X-ray scattering from crystals one sometimes uses an antisymmetric function in which, to use a metaphor, peaks are antisymmetrically related to nearby valleys.

I have often believed that my colleagues and other friends in the arts and humanities could appreciate the beauty of mathematics and science. Therefore, I shall try to illustrate abstract symmetry. Consider the three dimensional symmetry of a square, which has elements of symmetry: a four-fold axis, four two-fold axes lying in the plane of the square, a center of symmetry, a plane of symmetry in the plane of the square, and four planes of symmetry perpendicular to the plane of the square. Perhaps this is too much symmetry to keep in mind, so let me reduce it. Let me attach four identical right handed objects, one to each corner of the square in such a way that only the four fold axis remains.[20] Now I make an abstraction. I ask, what mathematical symbols will behave like a four-fold axis? The operations are rotations by $0°$, $90°$,

180°, 270°, after which one repeats. If we suppose that the rule of combination of the symbols is multiplication, we must find the symbol 1 after the four consecutive 90° rotations because we are back to our starting orientation. Hence $\sqrt[4]{1}$ is the appropriate choice. The choice of numbers is then limited to the four roots, 1, -1, $\sqrt{-1}$, or $-\sqrt{-1}$, any of which will represent the 90° rotation. Hence the four sets of numbers for the consecutive rotations of 0°, 90°, 180° and 270° are

0°	90°	180°	270°
1	1	1	1
1	-1	1	-1
1	$\sqrt{-1}$	-1	$-\sqrt{-1}$
1	$-\sqrt{-1}$	-1	$\sqrt{-1}$

Each one of these four horizontal lines is called a representation of the four fold axis. By squaring the symbol for the 90° rotation one finds the symbol for the 180° rotation. Similarly, the cube of the symbol listed for 90° gives that for 270°. This abstraction from a geometrical shape to pure mathematics is beautiful in its own right, but the next step is remarkable. These numbers can then be used to derive altogether different properties of a molecule which has a four fold axis (3). Examples of these properties are the nuclear spin orientations, the molecular vibrations, or the molecular wave functions. Almost every one of the last 20 years, I have taught a course in which this type of mathematics of symmetry is applied to molecular properties. Normally, it takes several weeks of lectures and problems before most students suddenly perceive these abstract relationships, and when they do, they recognize it as an aesthetic experience.

When the performance of two consecutive symmetry operations gives a different result if the operations are performed in reverse order, it is no longer possible to use numbers for the representations. This is the case for the four fold 90°

13

operation (C_4^1) followed or preceded by one of the two fold rotations (C_2) about an axis perpendicular to the four fold axis. The reader may see from a diagram that the order of the pair makes a difference: $C_4^1 C_2 = C_2 C_4^1$. Here two by two matrices are required to abstract these symmetry elements.

As we take leave of the symmetries of two- and three-dimensional space, we come gradually to totally unfamiliar behavior. The wave function of the electron is not left unchanged by a 360° rotation; its invariance requires a 720° rotation! [21] In the internal space SU(2) which describes the spin of the electron there is a two-valued representation by 2 x 2 matrix rotations which maps onto a three dimensional rotation through 720°. A plausible, although inexact, analogy illustrates the point. A line which is traced on the surface of a circular mobius strip repeats after 720°, not after 360°.

Symmetry in science has such a fundamental place that one is filled with wonder at the varieties of nature which can be described. I still remember the wonderful experience upon learning that symmetry of translation in space leads to the law of conservation of momentum, symmetry of orientation in space to conservation of angular momentum, and symmetry of time translation to the law of conservation of energy.[22] Another kind of symmetry arises because we cannot permanently label elementary particles: the description of their behavior has to be invariant upon interchange of any two identical particles. Then, either the wave function of a system containing these particles changes sign (for odd half integer spin) or does not change sign (for even half integer spin). These results can be related in part to time reversal symmetry.[23] The form in which this interchange symmetry is displayed shows that two electrons with parallel spins cannot exist in the same single atomic or molecular energy level. Recent studies of symmetries have related these two types of interchange symmetry in a kind of supersymmetry.[24,25,26]

14

The particles of physics and their associated fields show complex and subtle symmetries of unfamiliar types. The operations of charge conjugation (e.g. electron, positron [27] C, of parity space inversion symmetry) P and time reversal symmetry T are related by a very general invariance of the product PCT. It was not realized before the work of Yang and Lee that Nature makes a distinction between right and left at the level of elementary particles. In particular P is not conserved in the weak interactions. The violation of the product CP, implying failure of time reversal symmetry, is known only for the K° - $\bar{K}^{\circ}$ meson system. There may be some relationship between these violations and one of the grand asymmetries of the universe: the existence of matter, and of almost no anti-matter.[28,29] There are several models of this asymmetry. Weinberg's proposal, for example, has a departure from thermal equilibrium in the early stages of the expansion (T° ∿ 10^{28}K) of the universe during which a proposed heavy boson X (which mediates baryon non-conservation) decays in at least two different ways, violating CP symmetry.[30]

The asymmetry of time, as exhibited by spontaneous processes which involve matter and radiation, is clearly indicated in systems which are partially isolated from the rest of the universe. And yet, except for the implication of the K° - $\bar{K}^{\circ}$ meson system, the laws of particle physics are reversible in time. Asymmetry of time is also described in thermodynamics by coarse graining. This procedure defines most probable variables, such as pressure, temperature and entropy, by neglecting the complex correlations of individual particles. However, even if these correlations were included their weak coupling to the rest of the universe would tend to average the find grained structure. Ultimately then, asymmetry of time is most probably related to the irreversible part of the expansion of the universe, particularly in its early history.

Of course, you know that time moves from the past to the future, and that you remember the past but not the future.

The only exception is Lewis Carroll's White Queen: "It's a poor sort of memory that only works backwards." (5). It is challenging enough to show students how the laws of irreversible thermodynamics go forward in time, inasmuch as they are based on physical laws that are reversible in time. These methods which succeed for chosen parts of the universe need complete restudy when applied to the universe as a whole. In the very early stages of the universe, the nature of time, space and causality may be different from our present concepts of these properties. It is unknown whether these properties of the universe arose from total symmetry, although it is difficult to imagine otherwise.

Einstein's discovery that the physical laws are the same for two observers in a state of relative motion is basically a symmetry law. In particular, Maxwell's equations, which united electric and magnetic forces, were discovered to be invariant under a four coordinate (Lorentz) transformation; this, Minkowski showed, was a kind of rotation in a four-dimensional space in which the special character of time is taken into account. When generalized, and added to the equivalence principle, this transformation led Einstein to the general theory of relativity. These ideas originated field theories, in which symmetry principles control interactions.

But symmetry considerations became even more dominating as quantum mechanics developed, primarily because of the phase (for stationary states: the sign) of the wave functions which describe a system. For example, any single wave function for the isolated hydrogen atom must preserve the equivalence of x, y and z directions. If some solution has a direction, say x, then there must be equivalent solutions of the same energy associated with y and z, again to preserve directional equivalence. The orthogonality of x, y and z has its generalization in the infinite number of stationary states, each with its own symmetry characteristics.

The proton and neutron, which have nearly the same mass obey an approximate symmetry known as "isotopic spin" symmetry, which is like a kind of rotational symmetry except that it is the electric charge, for instance, of the particle which is altered, not the spatial orientation. This is a universal symmetry because the invariance property is maintained only if all protons and neutrons in the universe are transformed simultaneously.

When the field equations which describe a family of particles do not change when the symmetry transformation is made on some labeled property at each point of space and time, we have local invariance. Sometimes one begins with field equations which have a global but not local symmetry. Other fields are added to the theory in order to make the field locally as well as globally invariant. Such a theory is called a gauge theory. Symmetry can be broken when the solution adopted by Nature is less symmetrical than the original equations. (For example, the equations describing ferromagnetism show no preference in three dimensional space, but the magnet itself does adopt a particular direction of magnetization.) The breaking of a gauge symmetry may generate mass in a (virtual) particle which mediates forces between the particles of the field, and hence the underlying symmetry may be extremely difficult to recognize. This is the case for the photon and the Z° particle of the 1967 theory of Weinberg and Salam which unified electromagnetic and weak interactions. This theory, and the (unbroken) color symmetry theory of the strong interactions are progressing toward unification, and it is even possible that the last of the four presently known fields, gravity, will be included at least in its symmetry aspects.

These ideas of symmetry also introduce great simplicity [31] where there was complexity. They hint at a connection between the smallest particles and the origin of the whole universe and leave us feeling that we have only begun to appreciate the universe in which we live.

Relation to Aesthetics

It is not necessary to have followed all of these examples of symmetry in order to appreciate that the ideas are far more subtle, more general, and more beautiful than what one usually perceives as symmetry, balance and its deviations in the visual arts. The mastery of the language of group theory which describes these symmetries in an abstract way is itself an aesthetic experience available to those who will make the effort. Actually, one must make no less of an effort in order to appreciate in full the music and art of the present century. For example, my aesthetic appreciation is far greater if I become involved in a performance of a chamber music composition, and would, I expect, be even greater if I were capable of composing it.

We receive, during our lifetime an incredible amount of information through our senses, all of which requires selection, ordering, processing and abstraction. An example is the storage of vast amounts of language and meaning in ordered rules. We process this information continually both logically and intuitively, and we assimilate new information, often with revision of the whole or part of the structure. From particular events we distill the essence of things, ideas, hypotheses, and carry these one step further, sometimes testing against experience. However, the most searching tests are against the other ideas in the mind. When we do this consciously, we are not usually surprised, especially if we can trace the deductive or inductive processes. But when the subconscious mind presents us with a solution, and we are unaware of how this came about, we may be surprised. I think that this happens in our subconscious continually when we are awake, and probably to some extent in the free associations experienced in dreams. Of course we reject, or tend not to form, those associations which are inconsistent with experience, but occasionally an alternative and completely different set of ordered rules, of which we may not even be aware, will suddenly form and accommodate a group of ideas. When this happens one finds great aesthetic pleasure in a creative situation.

For an observer, aesthetic pleasure can be found if the artist has conveyed a new way of ordering ideas, or if the artist has supplied the missing part that completes the observer's own creative experience. It is in this sense that the aesthetic experience becomes universal. Creative individuals are repeatedly so, and they respond especially when a beautiful result is one that they almost obtained. If their response is not a selfish one, it can be an especially intense aesthetic experience as an observer.

Two matters remain. The first is that so little of these matters are accessible in our schools and universities. If one actually set out to give as little help as possible to both aesthetics and originality in science, one could hardly devise a better plan than our educational system. For example, one rarely hears about what we do not understand in science, and least of all how to prepare for creative ideas. Fortunately, the schools that I attended made so little demands that I had time for my own independent study of science and music. Actually, the most important part of my formal early education was the precise and accurate use of English language: its decline may be the greatest fault of education today. Of course, there are some wonderful exceptions, and for me these have been L. F. Jones (my high school science teacher), Linus Pauling, and my students and colleagues over the years.

The last matter is how to communicate the systematic body of knowledge that is developing in the area of aesthetics of science.[32] Without leisure, space in journals, or closeness of teaching and research, the aesthetic aspects are lost, almost preferentially. A new discovery in science, like a work of art, is almost immediately taken from its context, isolated by admiration, and removed from its human connections so that it is no longer part of life (1). When I fail to supply these connections, I find that the response to my new discovery is that this is one more thing to be learned. As a body of aesthetics in science grows, let us hope that it remains flexible enough to accommodate the surprises of the future. In the area of publication, one can hope that aesthetic criteria may also be used, even though a work may be incomplete or even incorrect. I do believe this was understood by the late Samuel Goudschmidt, but he was a remarkable exception. On the question of whether scientists should use aesthetic criteria in their research and choice of problem, my observation is that we already do so, quite naturally. We have, for example, employed the aesthetic concepts of symmetry and simplicity to great advantage because our minds work that way, and to a surprising extent so does nature. Perhaps this is because we are part of nature.

Let me close with a little history: Surrounding the heads of Roman emperors, Greek gods, and religious figures of Christianity and the far east, one sometimes sees concentric rings of color. It is likely that this is the artist's rendition of the glory. This is a name given to colored rings from sunlight refracted back toward the observer who looks toward the projection of his shadow into a fog or low cloud. In 1874 the physicist C.T.R. Wilson became so excited by this wonderful phenomenon that he built a cloud chamber in order to recreate the glory in his laboratory. However, he became distracted by the observation of visible tracks which were produced by energetic charged particles as they passed through the cloud in his apparatus. Wilson received the Nobel Prize for this work in 1927. I have seen the glory effect, and have made a Wilson cloud chamber when I was a youth.

Both effects are beautiful indeed. I can imagine the response upon seeing either of these effects for the first time!

Finally, I hope that I have shown you not only some of the aesthetic aspects of science, but also the human side, the intellectual and emotional breadth, and the artistic side of science as well.

(1) "For only when an organism shares in the ordered relations of its environment does it secure the stability essential to living. And when the participation comes after a phase of disruption and conflict, it bears within itself the germs of a consummation akin to the aesthetic." John Dewey, *Art as Experience,* Milton, Balch and Co., New York, 1934. Chapter 1.

(2) Galois' ideas were summarized in a letter to a friend on the eve of his death. "This letter, if judged by the novelty and profundity of ideas it contains, is perhaps the most substantial piece of writing in the whole literature of mankind." H. Weyl, *Symmetry,* Princeton University Press, Princeton, N.J., 1952.

(3) Mathematicians are like Frenchmen: whatever you say to them they translate into their own language and forthwith it is something entirely different (Goethe). Similarly in art: "I try to show that when some things are taken out of the usual context and put in the new, they are given an entirely new character." (Romare Beardon, in acknowledging his debt to Vermeer and Delacroix).

(4) "After months, you look at a problem from a different point of view and suddenly it breaks apart and seems that it wasn't difficult at all. There's something awe-inspiring. You aren't creating. You're discovering what was there all the time, and that is much more beautiful than anything that man can create." (Charles Fefferman, on the occasion of the Fields Medal, 1978.)

21

(5) "Through the Looking Glass," by Lewis Carroll.

(6) S. Chandrasekhar *(Physics Today,* July 1979, p. 25) attributes this definition of beauty to Heisenberg's essay, "The Meaning of Beauty in the Exact Sciences," Francis Bacon, "There is no excellent beauty that hath not some strangeness in the proportion!"

(7) P. Klee, "On Modern Art," Faber and Faber, London and Boston, 1948, p. 13.

"May I use a simile, the simile of the tree? The artist has studied this world of variety and has, we may suppose, unobtrusively found his way in it. His sense of direction has brought order into the passing stream of image and experience. This sense of direction in nature and life, this branching and spreading array, I shall compare with the root of the tree.

"From the root the sap flows to the artist, flows through him, flows to his eye.

"Thus he stands as the trunk of the tree.

"Battered and stirred by the strength of the flow, he moulds his vision into his work.

"As, in full view of the world, the crown of the tree unfolds and spreads in time and in space, so with his work.

"Nobody would affirm that the tree grows its crown in the image of its root. Between above and below can be no mirrored reflection. It is obvious that different functions expanding in different elements must produce vital divergences."

(8) S. Chandrasekhar, *Physics Today,* July 1979, p. 25.

(9) The conflict with Keats, "Beauty is truth, truth beauty— that is all Ye know on earth, and all Ye need to know," is only apparent: the relationship refers to the lore of good and evil in the wisdom of man, not to the meaning of truth in science. This analysis appears in Reference 1, p. 34.

(10) See the remarkable extension to include quantum gravitational aspects, by S. W. Hawking, *Nature, 248,* 30 (1974).

(11) C.-N. Yang, *Physics Today,* June 1980, p. 42.

(12) *A. Einstein: Philosopher-Scientist,* Vol. 1, p. 45. Pen Court Classics, p. 116. Ed: P. A. Schlipp, Third Ed. 1969.

(13) Alice: "How would you like to live in Looking-Glass House, Kitty? I wonder if they'd give you milk in there? Perhaps Looking-Glass milk isn't good to drink." Lewis Carroll, *Through the Looking-Glass.*

(14) S. L. Miller and H. C. Urey, *Science 117,* 528 (1953). The atmosphere in these experiments was a mixture of water, methane, ammonia and hydrogen.

(15) Private communication. A detailed structure of a right handed B-DNA fragment with the self-complementary sequenced (CpGpCpGpApApTpTpCpGpCpG) has been completed by R. Wing, H. Drew, T. Takano, C. Broka, S. Tanaka, K. Itakura and R. E. Dickerson, *Nature* 1980 in press.

(16) An English description has been given by W. N. Lipscomb in *Technique of Organic Chemistry,* Vol. 1 (Third Edition) Physical Methods Part II (A. Weissberger, ed.) 1960, p. 1641.

(17) *The Search for Solutions,* H. F. Judson; Holt, Rinehart and Winston, New York, 1980, p. 40.

(18) A. V. Shubnikov and V. A. Koptsik, *Symmetry in Science and Art,* Plenum Press, New York 1974.

(19) A. L. Loeb, *Color and Symmetry,* Wiley Interscience (1971).

(20) An N fold axis consists of consecutive rotations by 360°/N, until one repeats after N steps.

(21) S. A. Werner, R. Colella, A. W. Overhauser and C. F. Eagen, *Phys. Rev. Lett., 35,* 1053 (1975).

(22) E. L. Hill, *Rev. Mod. Phys., 23,* 253 (1951).

(23) J. Schwinger, *Phys. Rev., 82,* 914 (1951).

(24) D. Z. Freedman and P. van Nieuwenhuizen, *Sci. Am., 238,* p. 126, Feb. 1978.

(25) G. t'Hooft, *Sci. Am., 242,* p. 104, June 1980.

(26) H. Georgi and S. L. Glashow, *Physics Today 33,* p. 30, Sept. 1980.

(27) Indeed, the discovery of Dirac (1928-1930) that a relativistic theory of the electron admitted an equally valid solution for the then undiscovered positron is an extraordinary recognition of a symmetry of nature, and a remarkable faith that nature must exhibit the alternative solution.

(28) M. S. Turner and D. M. Schramm, *Physics Today*, 32, p. 42, Sept. 1979.

(29) G. Steigman, *Ann. Rev. Astron. Astrophys.*, *14*, 339 (1976).

(30) S. Weinberg, *Phys. Rev. Lett.*, *42*, 850 (1979).

(31) S. Weinberg, *The Nature of the Physical Universe, 1976 Nobel Conference,* D. Huff and O. Prewitt, Editors, 1979, John Wiley & Sons, p. 47.

(32) *On Aesthetics in Science,* J. Wechsler ed., MIT Press, Cambridge, Massachusetts 1979.

Beauty and Theoretical Physics

Chen Ning Yang

That there is beauty in science is a feeling shared by all scientists. What is the definition of beauty? The *Webster Collegiate Dictionary* defines beauty as "The quality or aggregate of qualities in a person or thing that gives pleasure to the senses or pleasurably exalts the mind or spirit." That is a good definition in less than 25 words. But, of course, the concept of beauty is actually more complex than that. You could ask what is meant by beauty in literature, or in a painting, or in music, or in the sciences? When you ask these questions you realize that the problem is complex indeed, and that an adequate definition can perhaps never be given.

But that scientists have known from early on that there is great beauty in science is beyond question. The first sentence of Copernicus' great book, *The Revolutions of the Heavenly Spheres,* published in 1542, reads: "Among the many and varied literary and artistic studies upon which the natural talents of man are nourished, I think that those above all should be embraced and pursued with the most loving care which have to do with things that are very beautiful." That he chose this sentence to begin his book is a clear indication of how much he appreciated the aesthetic dimension of science. In fact, throughout the book one finds similar passages expressing unbelievable pleasures of the mind.

Now the beauties of the different scientific disciplines are deeply related, but they are not quite the same. My topic is beauty in theoretical physics, and, as Professor Lipscomb has already emphasized, a lot of the beauty in the physical sciences—and certainly in theoretical physics—is deeply related to the concept of beauty in mathematics. Maxwell, one of the greatest physicists of the nineteenth century, in an 1870 presidential address to Section A of the Mathematical and Physical Sciences divisions of the British Association, spoke about his impressions of Professor Sylvester, a distinguished mathematician and his predecessor as the president. Maxwell said of Sylvester that he had exhibited the "idea of harmony of appreciation which he feels to be the root of all knowledge, the formation of all pleasures and the conditions of all actions. The mathematician has above all things an eye for symmetry."

In the same vein we find another great physicist of the nineteenth century, Boltzmann, saying:

> Even as a musician can recognize his Mozart, Beethoven, or Schubert after hearing the first few bars, so can a mathematician recogize his Cauchy, Gauss, Jacobi, Helmholtz or Kirchhoff after the first few pages. The French writers reveal themselves by their extreme formal elegance, while the English, especially Maxwell, by their dramatic sense. Who, for example, is not familiar with Maxwell's memoirs on his dynamical theory of gases? . . . The variations of the velocities are, at first, developed majestically: then from one side enters the equations of state: and from the other side, the equations of motion in a central field. Ever higher soars the chaos of formulae.

26

Suddenly, we hear, as from kettle drums, the four beats 'Put n = 5.' The evil spirit V (the relative velocity of the two molecules) vanishes: and, even as in music a hitherto dominating figure in the bass is suddenly silenced, that which had seemed insuperable has been overcome as if by a stroke of magic. . . . This is not the time to ask why this or that substitution. If you are not swept along with the development, lay aside the paper. Maxwell does not write programme music with explanatory notes. . . . One result after another follows in quick succession till at last, as the unexpected climax, we arrive at the conditions for thermal equilibrium together with the expressions for the transport coefficients. The curtain then falls!

So Boltzmann compares the pleasures that one obtains in reading Maxwell's great article on the dynamical theory of gases with the pleasure obtained in listening to great music.

In preparing for this conference I thought about the possibility of trying to define beauty in the sciences by some collection of words. Clearly, such words as "harmony," "elegance," "unity," "simplicity," "order," and so on, all have something to do with beauty in science and with beauty in theoretical physics in particular. But thinking about how to put these words together to form a definition of "beauty," I began to realize that, in fact, the concept of beauty in physics is not fixed. It has evolved because the subject matter of theoretical physics has evolved, and I strongly suspect that the same is true of all the sub-disciplines of science.

A most conspicuous and important aspect which affects our changing perception of the beautiful in theoretical physics is its increasing mathematization. Let me try to explain to you what I mean from a historical perspective. Galileo's time was universally regarded as the birthplace of modern physics. He left us a number of important discoveries, but if you think

27

about it, important as these discoveries are, they were not the most important heritage. It was, rather, that he taught us how to do physics.

Perhaps this is best illustrated by another passage from Maxwell's 1870 lecture:

> The feature which presents itself most forcibly to the untrained inquirer may not be that which is considered most fundamental by the experienced man of science: for the success of any physical investigation depends on the judicious selection of what is to be observed as of primary importance, combined with a voluntary abstraction of the mind from those features which, however attractive they appear, we are not yet sufficiently advanced in science to investigate with profit.

It was Galileo who taught the world of science the lesson that you must make a selection, and if you judiciously select the things that you observe, you will find that the purified, idealized experiments of nature result in physical laws which can be described in precise mathematical terms. That is the truly great lesson of Galileo, and that, of course also introduced the beginning of the quantitative science of physics. Galileo's was a profound and beautiful idea.

Next came Newton. We all know that Newton was the one who gave us a complete system of classical mechanics. It so dominated the science of physics for something like 200 years that it later became very difficult for people to make deviations from it. Through his work, the mathematization of physics progressed and became recognized as very rigorously mathematical.

Then came the latter part of the eighteenth century when physicists turned their attention to electricity and magnetism. In approximately 50 years time through great experimental studies, physicists learned the four fundamental laws of electricity and magnetism. The first was Coulomb's law. Coulomb's equipment consisted of a large sphere that could take a charge with a rod suspended next to it. Depending on whether the sphere was charged or not, the suspended rod next to it would alternate with a different frequency. Observing the frequency and the distance between the charged sphere and this rod, he was able to derive the Inverse Square Law, the first important law of the four laws of electricity and magnetism: Coulomb's law, Gauss's law, Ampere's law, and Faraday's law. These discoveries led to the beginning of the science of electricity and indeed led to the modern world.

But to make further progress we needed more than these empirical laws. We needed Maxwell's great concept of the field which was required to replace the concept of action at a distance—the prevalent concept before Maxwell. For instance, even a great mathematician and physicist like Gauss much preferred the concept of action at a distance. Faraday, on the other hand, felt that transmission of forces at a distance was incoherent; he preferred to say that each body acts on its immediate neighbor, which again acts on its neighbor, and so on, until eventually the force is transmitted to the other end.

The concept of lines of force and the concept of the field were great concepts. They were great not only because they led to a deepened understanding of electricity and magnetism.

However, great as these ideas of Faraday's were, they were intuitive ideas which were not put into mathematical form. Faraday had a great intuition, but he was not a mathematician. In fact, if you read his articles, you can hardly find a single mathematical formula. It is amazing how, without mathematical apparatus, he could have had such a penetrating intuition.

Maxwell, a man 40 years Faraday's junior, came along at exactly the right moment. Maxwell was a great admirer of

Faraday, and he tried to write down Faraday's ideas in mathematical form. This proved difficult precisely because Faraday's ideas were intuitive and not mathematical. Struggling for almost 20 years, he finally arrived at the great Maxwell's equations. The article in which Maxwell first formulated his equations is called "A Dynamical Theory of the Electromagnetic Field"—undoubtedly the greatest theoretical physics paper in the nineteenth century. Maxwell's equations represent the experimentally discovered Coulomb's law, the Gauss law, the Ampere law, and Faraday's law. The mathematical formulation of these laws was a great advantage. Putting them into mathematical form, Maxwell could bring to bear all the wisdom and the the technology that had been developed by the mathematicians over centuries. As a consequence he and his followers were able to manipulate the equations to derive results which otherwise would have escaped their grasp. Out of these equations Maxwell was able to conclude, for instance, that the science of light is nothing but a branch of electricity and magnetism. He was then able to predict that electricity and magnetism can propagate in waves. And these waves were indeed discovered ten years after Maxwell's death.

It would, however, be wrong to think that there was universal acceptance of Maxwell's endeavors and results. Faraday himself was quite afraid of this exercise of Maxwell's. In 1857, before Maxwell was completely finished with his work, he communicated some of his partial results to Faraday. Faraday responded with an interesting letter:

My Dear Sir,

I received your paper, and thank you very much for it. I do not say I venture to thank you for what you have said about "lines of force," because I know you have done it for the interests of philosophical truth: but you must suppose it is work grateful to me, and gives me

30

much encouragement to think on. I was at first almost frightened when I saw such mathematical force made to bear upon the subject and then wondered to see that the subject stood it so well.

The suspicion of experimental physicists—and some theoretical physicists—regarding complicated mathematical formalism is a very common one, and I would say it's healthy. The reason is because most of the complicated mathematical formalism of physical phenomena eventually have led to nothing. Nevertheless, while that is true, we must recognize greatness when greatness occurs and in this case when Maxwell was finished there was no doubt that the science of electricity and magnetism had turned a most important corner. Maxwell's equations led to an understanding of electric, magnetic and optical phenomena in great detail, in great accuracy, and in great completeness. Maxwell's equations led to all the subsequent developments of electrical engineering and the whole communication industry. On the purely academic side, the study of Maxwell's equations led to two of the profound revolutions of physics in the beginning years of this century: the special theory of relativity and the general theory of relativity, subjects I shall return to later.

In more recent times, the study of Maxwell's equations and their generalizations led to our assumption that the symmetry underlying physical laws represents one of the most important aspects of how nature is constructed. All of the interacting forces, all the forces of nature, follow from some fundamental forces, which are four in type. It is now generally recognized that the existence and properties of these fundamental forces all can be derived from some specific symmetry principles. All these developments were direct or indirect consequences of a deeper understanding of Maxwell's equations.

Mathematization is accelerating in recent physics. The special theory of relativity is based on the concept of the four-dimensional continuum of space-time. The general theory of relativity is based on Riemannian geometry, as Professor Lipscomb already mentioned. The concept of quantum mechanics is based mathematically on a beautiful and abstract mathematical theory called Hilbert space; and the physics of the non-abelian gauge theory is amazingly based on fibre bundle geometry which was discussed by the mathematicians independently of anything which has to do with physics starting in the 1920's and 1930's. All these mathematical developments which are of basic importance to the physics of the twentieth century are quite abstract. And they are very beautiful.

Perhaps we can now begin to see why this accelerated mathematization of physics leads to a change of what is regarded as beautiful in theoretical physics. Thinking about this I am led to suggest three categories of beauty: the beauty of phenomena, the beauty of theoretical description, and the beauty of the structure of theory. Of course, like all discussions of this type these are not sharply different types of beauties; they overlap, and there are beautiful developments which one finds difficult to put into any one category. But I would say as a general rule there are different types of beauties in theoretical physics, and our appreciation of these beauties is somewhat different depending on which special category we are talking about. Furthermore, as time goes on, our appreciations of the beauties of the different classes also have changing weights.

What do I mean by "beauty of phenomena?" Well, that's easy to explain. There are many physical phenomena which are beautiful to our immediate senses. When as a child we see the rainbow, we immediately would say that it's very beautiful. And of course there are beautiful experimental phenomena which require more training to observe, for example, that the planetary orbits are all ellipses—a very beautiful phenomenon. When it was first discovered that these were perfect ellipses there was great joy. Or take another example—the spectral lines. The spectral lines of atoms are very distinct and sharply defined aspects of optics. Maxwell, for one, found them to be beautiful because they seemed to be independent of the conditions in which you put the emitting atoms. You put the atoms under high pressure, and as nineteenth century scientists discovered, the spectral lines do not change at all. They seem to be exhibiting some intrinsic properties of the atoms; that, of course, is a very beautiful idea.

Or to give you another example: the phenomenon of superconductivity. Imagine the astonishment of the discoverer of this phenomenon when he found that you can have a coil carrying a current which will—without batteries—go on for months or years without winding down. So there are clearly beauties in physical phenomena.

What do I mean by "the beauty of theoretical description?" The law of the Coulomb force is a beautiful description; it's a description of phenomena which *a priori* do not have to obey any specific laws, and yet they do. The first and second laws of thermodynamics are beautiful theoretical descriptions of some fundamental properties of nature. The consequences of the first and second laws, and the precise observation of these laws, are subjects that every student of thermodynamics would appreciate.

To take another example, in the beginning of this century radioactivity was discovered. It was rapidly determined that radioactivity would cause the elements that are radioactive to disintegrate. But it was Lord Rutherford who gave us the precise law. It is an exponential decay law of great accuracy, and even today we have found no deviations from it.

Finally, what about "the beauty of the structure of theory?" When a theory is formulated, especially in the twentieth century, it has a tendency to have an elegant structure, usually a

33

mathematical structure of its own. That nature chooses such mathematical structures for its physical laws is a marvel that nobody has really explained. It is clear that the beauty of these mathematical ideas is a kind of beauty which is quite different from the other kinds that we have been talking about. And the increasing mathematization of physics means that this last aspect of beauty is becoming daily more important in our field.

Perhaps another example would be useful at this juncture: the deepening appreciation of the beauty of the periodic table. As you all know, the periodic table was first constructed in the last century when it was found that if you put columns of elements of similar properties together you have a nice table—though with gaping holes. This allowed people to look for the missing elements. One by one, they were found. This was an elegant result having great practical consequences as well. I would say, however, that it belongs to the beauty of phenomena. But then came the Bohr atom and quantum mechanics. These developments provided a more fundamental theoretical understanding of the structure of the periodic table, namely, that the place that an element should occupy in the periodic table is related to the number of electrons it carries in its atomic structure. This was a truly profound discovery.

Then, after quantum mechanics, our understanding was deepened once again. As Professor Lipscomb has mentioned, the periodic table involves periods of such lengths as 2, 2, and 6. These numbers are deeply related to a mathematical concept called "group theory" which describes the underlying symmetry of physical law. It is extraordinary that the fundamental symmetry concept, when put into the deep, profound mathematical language of group theory has led to these numbers in an unambiguous and unerring way. Scientists learned through such developments as this that nature has previously unimagined patterns that man can aspire to comprehend.

In recent physics we find quite frequently that we first derive an equation and then discuss the physical implications. This is an important departure from earlier patterns of development. We said before that the great laws of electricity and magnetism were experimentally discovered and then translated into mathematical form. After these equations were written down, people used them to find the underlying symmetry properties of electricity and magnetism which they expressed. In recent physics, on the other hand, one starts with the symmetry properties and arrives at the equations.

Historically, this departure from earlier patterns of development arose in the following way. When Maxwell's equations were first written down, they were regarded as highly obscure; people didn't understand them at all. In a memorial volume to H. A. Lorentz, Professor A. D. Fokker said:

Maxwell's writings were obscure and mysterious. For the generation which had to digest them the treatises on electricity and magnetism came to be, to quote Ehrenfest, a kind of intellectual jungle, all but impenetrable in its untamed fertility. Lorentz himself states: "It is not always easy to comprehend Maxwell's ideas. One feels a lack of unity in his book due to the fact that it recalls faithfully his gradual transition from old to new ideas." Lorentz, Heinrich, Hertz and Oliver Heaviside are the main elaborators of Maxwell's scientific inheritance. Lorentz here opens up great vistas.

What was Fokker referring to? Well Maxwell, great as he was, was not able to shake off the prejudices of his time. There was the question of whether there's a medium in which the electromagnetic field propagates. And if you read what Maxwell wrote, you will find that he was of two minds. Sometimes he seemed to say that this medium was not necessary, but sometimes he seemed to say that it was real. And this ambiguity in his own mind was transmitted to his writings causing great confusion.

In fact, the confusion, as already intimated, was only cleared up once and for all by Einstein in 1905. This 26-year-old physicist told the world that there was no medium; what people thought were merely mathematical games were real. Above all, Einstein had the courage to question the concept of simultaneity. That was perhaps the greatest stumbling block for all the physicists preceding him because everyone deeply felt that he understood the concept of simultaneity.

Two years later, Einstein decided to reverse his procedure because the symmetry which is deeply related to Einstein's concept of relative simultaneity was a consequence of the Maxwell's equation, as both Lorentz and Einstein pointed out. This was such an impressive development that Einstein decided to reverse what had been the normal pattern and start with a larger symmetry and then ask what equations could be derived as consequences that would preserve this symmetry. That was how the second revolution in twentieth century physics occurred.

Dirac wrote in May 1963 *Scientific American* that "it is more important to have beauty in one's equations than to have them fit the experiments." Dirac is the greatest living physicist. He was singularly able to perceive beauty where no one else had. To many physicists today what Dirac said contains a great truth. It is astonishing that sometimes if you follow the guidance toward beauty that your instincts provide, you arrive at profound truth, even though contradictory to experiments. Dirac himself was led in this way to the theory of anti-matter.

I will give you another example where the mathematical equations anticipated the physical data. It required a longer struggle and was a more complicated human story than Einstein's theory of special relativity, but it was an equally great development. Toward the end of the nineteenth century there

36

was a great debate in physics about whether thermodynamics—which is already founded in the mid-nineteenth century—was or was not based on an atomic and molecular theory of matter. It may sound very weird today that even toward the end of the nineteenth century there were people who did not believe there had to be atoms and molecules, but for many years some of the great physicists were of the opinion that the atomic and molecular theories were all wrong. For instance Boltzmann in his great book *Lectures on Gas Theory*, written in 1898, said:

In my opinion it would be a great tragedy for science if the theory of gases were temporarily thrown into oblivion because of a momentary hostile attitude toward it, as for example was the wave theory because of Newton's authority. I am conscious of being only one individual struggling weakly against the stream of time. But it still remains in my power to contribute in such a way that, when the theory of gases is again revived, not too much will have to be rediscovered.

What was the "stream of time" that he was talking about? It was the idea that the molecular theory of gases and the statistical basis of thermodynamics were completely wrong. Many of the very distinguished physicists and chemists of that time, including Professor Ostwald, who was a Nobel prize winner of 1909, were deadset against the molecular theory. And that led to such profound despair on the part of Boltzmann that, together with some of his other difficulties, he fell into a deep depression and finally committed suicide in Italy in 1906.

Nor was Boltzmann the only person who suffered in this respect. Gibbs, the greatest American physicist of the nineteenth century, also had to struggle with the same problem. At that time there were a variety of experimental results contradicting the molecular theory, and that is part of the reason why the forces against the theory had the upper hand. Yet Gibbs pursued his study of what he called "the rational foundation of thermodynamics." In 1902 Gibbs wrote a book called *Elementary Principles in Statistical Mechanics, Developed with Special Reference to the Rational Foundation of Thermodynamics.* Together with Boltzmann's ideas, this book founded the science of statistical mechanics. He says there:

> Moreover, we avoid the gravest difficulties when, giving up the attempt to frame hypotheses concerning the constitution of material bodies, we pursue statistical inquiries as a branch of rational mechanics. In the present state of science, it seems hardly possible to frame a dynamic theory of molecular action which shall embrace the phenomena of thermodynamics, or radiation, and of the electrical manifestations which accompany the union of atoms. Yet any theory is obviously inadequate which does not take account of all these phenomena. Even if we confine our attention to the phenomena distinctively thermodynamic, we do not escape difficulties in as simple a matter as the number of degrees of freedom of a diatomic gas. It is well known that while theory would assign to the gas six degrees of freedom per molecule, in our experiments on specific heat we cannot account for more than five. Certainly, one is building on an insecure foundation, who rests his work on hypotheses concerning the constitution of matter.

And of course we know what happened in later history. By 1925 when quantum mechanics was invented it became clear that the disagreement with experiments was not the fault of Boltzmann or Gibbs. It was the fault of the theory. Once the theory was replaced by quantum mechanics, quantum statistical mechanics was born according to Gibbs' "rational foundation." His findings turned out to be in complete agreement with all experimental results.

With all this emphasis in theoretical physics on beauty, it is perhaps not surprising to find that many great physicists today repeatedly emphasize the importance of beauty for future work in physics. Einstein in 1933 said that "the creative principle resides in mathematics. In a certain sense, therefore, I hold it true that pure thought can grasp reality, as the ancients dreamed." Or, (in 1934) "The theoretical scientist is compelled in an increasing degree to be guided by purely mathematical, formal considerations." I had earlier quoted Dirac for you saying that if he had to choose between beauty and agreement with experiments he would choose beauty. In the case of both Einstein and Dirac, this emphasis is not surprising. If you look at the style of their work in physics, beauty has always been a guiding principle.

It is somewhat strange that Heisenberg also took the same view. In 1973, a few years before he died, Heisenberg said: "We will have to abandon the philosophy of Democritos and the concept of elementary particles. We should accept instead the concept of fundamental symmetries."

This may come as a surprise from Heisenberg, whose work is not distinguished by its quest for beauty as much as by its insistence on correlation of fact and experiment. Heisenberg, we all know, made one of the greatest contributions in twentieth century physics when he was among the first to create the science of quantum mechanics and then was the first to state the Uncertainty Principle. But I still think it is correct to say that you do not find the same guidance by principles of beauty in Heisenberg's work that you find in either Dirac's or Einstein's. So how are we to understand the passage quoted above? I think actually it is not difficult to understand. In the same book Heisenberg also said about the development of quantum mechanics of 1925: "The new mathematical scheme is very different from the old one. The surprise was that such a scheme existed at all. Bohr had the impression before that time, that we know the Newtonian mechanics doesn't work, and that may mean that nature is so irrational that we can never get any consistent mathematical scheme for description." In other words, in 1925, before quantum mechanics was invented, things were so slippery, they were so strange, that Bohr, who was a towering figure, feared that nature might be irrational. Guided by his insistence on adherence to the experiments, Heisenberg wrote down some equations. He did not understand the mathematical structure of these equations, because he was just a young man. He had never learned matrix theory. So he wrote down the equations which Bohr, his mentor, recognized as matrix multiplications. With this in mind, you will understand him when he said, "Then we saw that mathematicians did things we couldn't do ourselves." So I would say that Heisenberg came to the appreciation of beauty in theoretical physics by a different route. He was not following the instincts of being guided by beautiful concepts. And he was puzzled by what he was seeing, when finally it turned out that his mathematics anticipated the facts completely. From that moment on he was a convert to mathematical beauty.

In conclusion, let me ask, what is the final criterion of what is beautiful? I think the answer depends on the particular area of study. In the natural sciences, I would argue that the final judge is whether it is utilized in nature. In this respect, beauty in the sciences is different from beauty in mathematics where the final criterion must be whether the beauty relates to the rest of mathematics. This may not be the criterion for mathematics in earlier centuries, but it is correct today. Finally, besides natural science and mathematics, other areas where beauty is important are art, literature and music. In these areas I would contend that the final criterion for what is beautiful is whether man relates to it.

Manchester and Athens

Freeman J. Dyson

I. As Great a Human Exploit

This title is taken from a novel published in 1844 by Benjamin Disraeli, at that time a young and flamboyant and not very successful politician. Thirty years later he became Prime Minister of Great Britain and chief architect of the imperial dreams which filled the second half of Queen Victoria's long reign. The novel [1] [Disraeli, 1844] is called, "Coningsby, or the New Generation," and it did for Disraeli roughly the same job that "Profiles in Courage" and "Why Not the Best?" did for John Kennedy and Jimmy Carter. It gave the voters a chance to see what kind of person Disraeli was. It showed them, among other things, that Disraeli was the kind of person who would find it natural to write his statement of political beliefs in the form of a romantic novel. It is a story about a young man starting a career in politics at a time of rapid economic and social change. I do not wish to claim that Disraeli was a better writer than Kennedy or Carter, but he certainly had a more vivid imagination, and perhaps also a deeper understanding of history. About half-way through the

41

story, Disraeli's hero spends a few days in Manchester, and here are the thoughts which the Manchester of 1844 called to his mind:

"A great city, whose image dwells in the memory of man, is the type of some great idea. Rome represents conquest; Faith hovers over the towers of Jerusalem; and Athens embodies the pre-eminent quality of the antique world, Art. . . . What Art was to the ancient world, Science is to the modern: the distinctive faculty. In the minds of men the useful has succeeded to the beautiful. Instead of the city of the Violet Crown, a Lancashire Village has expanded into a mighty region of factories and warehouses. Yet, rightly understood, Manchester is as great a human exploit as Athens.

"The inhabitants, indeed, are not so impressed with their idiosyncrasy as the countrymen of Pericles and Phidias. They do not fully comprehend the position which they occupy. It is the philosopher alone who can conceive the grandeur of Manchester, and the immensity of its future. There are yet great truths to tell, if we had either the courage to announce or the temper to receive them."

Disraeli, of course, knew very little about science. If he had had any real understanding of science, he would not have described the antithesis between Athens and Manchester as an antithesis between Art and Science. As a matter of historical fact, Athens was in ancient times preeminent in philosophy as well as in art and literature. The ancients made no distinction between science and philosophy. Athens was the city of Plato and Aristotle as well as of Pericles and Phidias. Aristotle deserves as well as anybody the title of chief scientist of the ancient world. The Athenian Academy was the original prototype of a professional scientific institution, and for several hundred years, until the rise of the Hellenistic civilization with its intellectual center at Alexandria, Athens could fairly claim to be the scientific capital of the world. There was never a time when anybody who knew anything about science could make such a claim for Manchester. In 1844, when Disraeli published "Coningsby," Manchester was a serious center of scientific activity, but not so serious as to eclipse London and Paris and Berlin. John Dalton, who first made the connection between the old Greek idea of atoms and the facts of quantitative chemistry, and James Joule, who made thermodynamics into an exact science, were both Manchester men. But it was not Dalton's atoms or Joule's experiments which excited Disraeli's imagination. When Disraeli wrote "Rightly understood, Manchester is as great a human exploit as Athens," he had something else in mind. If science had been the cause of Disraeli's excitement, he need not have gone to Manchester to find it. Michael Faraday was doing his electro-magnetic experiments, and giving his famous public lectures, right on Disraeli's doorstep in London.

What was so exciting about Manchester? Disraeli with his acute political and historical instinct understood that Manchester had done something unique and revolutionary. Only he was wrong to call it science. What Manchester had done was to invent the industrial revolution, a new style of life and work which began in that little country town about two hundred years ago and inexorably grew and spread out from there until it had turned the whole world upside-down. Disraeli was the first politician to take the industrial revolution seriously, seeing it in its historical context as a social awakening as important as the intellectual awakening that occurred in Athens 2300 years earlier. Disraeli saw it as his task to bring together the two worlds into which England was split, the old world of aristocratic glitter surrounding the Queen in London, the new world of factories and warehouses spreading out from Manchester. He succeeded, and the true memorial of his success is the name which all over the world belongs to the era in which he lived, the Victorian Era. Now, a hundred years later, the world is looking in vain for another Disraeli, a statesman who combines the vision of a historian with the manipulative skill of a master politician.

But Disraeli's political achievements, like the technical achievements of the early industrial revolution, had little to do with science. Disraeli was mistaken in regarding Manchester as a scientific enterprise. The truth of the matter was best expressed by L. J. Henderson: "Science owes more to the steam-engine than the steam-engine owes to science." Science did flourish in Manchester during the crucial formative years of the industrial revolution, but the relations between science and industry were not at all in accordance with Disraeli's ideas or with the ideas of later Marxist historians. Science did not arise in response to the needs of industrial production. The driving forces of the Manchester scientific renaissance were not technological and utilitarian; they were cultural and aesthetic.

Last year the historian Arnold Thackray was in Princeton and I had the opportunity to learn from him what really happened in Manchester in the second half of the eighteenth century. Thackray published his findings a few years ago in the American Historical Review in an article with the title "Natural Knowledge in Cultural Context: The Manchester Model," [Thackray, 1974]. It turns out that the seminal influences in the growth of science in Manchester were the city infirmary (founded in 1752) and the Cross Street Unitarian Chapel. The doctors and the unitarians were the intellectual élite of the rapidly growing town, and they joined forces in 1781 to create an institution appropriate to their needs, the Manchester Literary and Philosophical Society. The Literary and Philosophical Society was consciously designed to give Manchester a cultural focus, to raise the aspirations of the leading citizens to a loftier level, to divert them from the mere accumulation of wealth to the pursuit of higher learning. From the very beginning the society was enormously successful. It attracted and supported first-rate scientists such as Priestley, Dalton and Joule, it published a journal, it built a library and a College of Arts and Sciences and a Mechanics' Institution and an Academy, and it gave birth to Owens College which ultimately grew into the University of Manchester. During the first seventy years of its existence, from 1781 to 1851, the society had altogether 588 members, and of these no less than 31, a little over five per cent, became sufficiently distinguished as scientists on the national scene to be elected Fellows of the Royal Society of London. An extraordinary achievement for a bunch of amateurs in a raw provincial town.

The historical question which needs to be answered is why a group of doctors and unitarian chapel-goers, having decided to give their city a new cultural identity, should have found it in the study of physics and chemistry. The name of the Literary and Philosophical Society shows that their original objective was general culture and not specialized science. Thomas Henry, who was both a physician and a unitarian and one of the founding members of the society, expressed their purpose most explicitly: "A taste for polite literature, and the works of nature and art, is essentially necessary to form the gentleman," [Henry, 1785]. In other words, the founding fathers wanted to prove that it was possible to live in Manchester and still be a gentleman. How did it happen that the search for genteel status led them so rapidly and decisively into science?

Arnold Thackray explains their concentration upon science as a result of two main factors. First, the unitarians were legally barred from the academic establishment of Oxford and Cambridge, and the atmosphere of Manchester was saturated with contempt for the ancient universities, then at a low point of decadence and corruption. So the organizers of the Literary and Philosophical Society were anti-academic, having no use for the smattering of Latin and Greek which the universities in those days called a classical education. Second, the inhabitants of Manchester were unrepresented in Parliament and therefore tended to radicalism in politics, especially in the formative years of the society before the French Revolution made radicalism unpopular. Radical politics included a belief in public education, and science served better than Latin and Greek as a vehicle for educating the masses. The chemist Priestley, hero of the radicals, expressed directly their view of science as an agent of social reform: "The English Hierarchy, if there be anything unsound in its constitution, has reason to tremble even at an air pump or an electrical machine," [Priestley, 1774].

So the anti-academic, anti-establishment brashness of Manchester made a fertile ground for the growth of science. And the science which grew in that northern soil had a style very different from the science of Athens, just as two hundred years later the music of the Beatles growing up in nearby Liverpool had a style very different from the music of Mozart. It is difficult to describe a style in one word without over-simplification, but still I will try to do it. It seems to me that the essential difference between the science of Athens and the science of Manchester is that Athens is attempting to unify while Manchester is attempting to diversify. The science of Athens emphasizes ideas and theories; it tries to find unifying concepts which tie the universe together. The science of Manchester emphasizes facts and things; it tries to explore and extend our knowledge of nature's diversity. Of course the tradition of unifying science did not end with Athens, and the tradition of diversifying science did not begin with Manchester. Historians of science are accustomed to call these two traditions in science Cartesian and Baconian, since Descartes was the great unifier and Bacon the great diversifier at the birth of modern science in the seventeenth century. The unifying and diversifying traditions have always remained alive in science to a greater or lesser extent. But the human exploit which Disraeli discerned in Manchester included an important rebirth of the diversifying tradition in science. Manchester brought science out of the academies and gave it to the people. Manchester insolently repudiated the ancient prohibition: "Let nobody ignorant of geometry enter here," which Plato is said to have inscribed over the door of his academy in Athens.

II. Unifiers and Diversifiers

Let me now come to the modern era. The years 1907-1919 were the high noon of physics in Manchester. During those years Ernest Rutherford was professor in Manchester, at the peak of his youthful vigor, creating the science that later came to be called nuclear physics. In Manchester he discovered the atomic nucleus and observed the first nuclear reactions. He was a scientist in the tradition of the Manchester Literary and Philosophical Society, self-confident and largely self-taught, disrespectful of academic learning, more interested in facts than in theories. Concerning theoretical physicists he made the famous statement: "They play games with their symbols, but we turn out the real facts of Nature." [Blackett, 1954]. In 1913 the great theoretician Niels Bohr sent to Rutherford in Manchester the manuscript of his epoch-making paper on the quantum theory of the atom. Rutherford understood the importance of Bohr's work and volunteered to send the paper to the Philosophical Magazine for publication. But at the end of Rutherford's letter to Bohr there is a brief postscript: "I suppose you have no objection to my using my judgement to cut out any matter I may consider unnecessary in your paper? Please reply," [Bohr, 1961]. Rutherford stood in awe of nobody. Not even of Einstein.

The astronomer Chandrasekhar tells a story of a conversation which he heard in 1933 between Rutherford and Eddington [Chandrasekhar, 1979]. Eddington was the astronomer who obtained the crucial evidence of the correctness of Einstein's General Relativity by observing the gravitational deflection of light-rays by the sun at the total eclipse of 1919. A friend said to Rutherford in the presence of Eddington and Chandrasekhar, "I do not see why Einstein is accorded a greater public acclaim than you. After all, you invented the nuclear model of the atom, and that model provides the basis for all of physical science today. . . ." Rutherford, in response, turned to Eddington and said, "You are responsible for Einstein's fame." Rutherford was not belittling Einstein's achievement, but he went on to say that the typhoon of publicity surrounding Einstein was a result of the dramatic circumstances of the eclipse expeditions and of the dramatic manner in which Eddington's results were announced in London. Clearly, Rutherford felt that Einstein's fame was somewhat disproportionate, and he was not afraid to say so.

The point of telling this story is not to pull down Einstein from his position as the greatest scientist of our century. Chandrasekhar ends his account by remarking that the true and lasting preeminence of Einstein lies in the incredibly rich content of the general theory of relativity itself. Chandrasekhar agrees with Einstein's own triumphant statement in his first presentation of the theory to the Berlin Academy [Einstein, 1915]: "Scarcely anyone who has fully understood this theory can escape from its magic." The point of the story is that Rutherford did not understand general relativity and was therefore immune to its magic. Rutherford's science and Einstein's science were so different in style that no real understanding between them was possible. The gulf between them was deeper than the normal gulf between experimenters and theorists. The problem was not merely that Einstein did not care about alpha-particles and Rutherford did not care about curvature-tensors. The problem was that they had fundamentally different concepts of the nature and purpose of science. Einstein said: "The creative principle resides in mathematics.

In a certain sense, therefore, I hold it true that pure thought can grasp reality, as the ancients dreamed," [Einstein, 1933]. Rutherford said: "Continental people do not seem to be in the least interested to form a physical idea as a basis of Planck's theory. They are quite content to explain everything on a certain assumption and do not bother their heads about the real cause of a thing. I must say that the English point of view is much more physical and much to be preferred," [Blackett, 1954]. And on another occasion when Eddington remarked at the dinner-table that electrons were very useful conceptions but might not have any real existence, Rutherford replied, "Not exist, not exist,—why I can see the little beggars there in front of me as plainly as I can see that spoon," [Andrade, 1961].

So there are two styles of science typified by Athens and Manchester, Einstein and Rutherford, abstract and concrete, unifying and diversifying. The two styles are not in conflict with one another. They are complementary, giving us two views of the universe which are both valid but cannot both be seen simultaneously. The word "complementary" has here the technical meaning that it has in Niels Bohr's interpretation of quantum mechanics. According to Bohr, an electron cannot be pictured as a simple material object but must be described by two complementary pictures emphasizing its particle behavior and its wave behavior separately. Einstein and Rutherford gave us complementary views of science, and each was too single-mindedly attached to his own view to understand the other. Both of them, for opposite reasons, rejected the compromise which Bohr's notion of complementarity offered them. For Einstein, the electron must ultimately be understood as a clumping of waves in a non-linear field theory. For Rutherford, the electron remained a particle, a little beggar that he could see in front of him as plainly as a spoon.

The discoveries of recent decades in particle physics have

49

led us to place great emphasis on the concept of broken symmetry. The development of the universe from its earliest beginnings is regarded as a succession of symmetry-breakings. As it emerges from the moment of creation in the big bang, the universe is completely symmetrical and featureless. As it cools to lower and lower temperatures, it breaks one symmetry after another, allowing more and more diversity of structure to come into existence. The phenomenon of life also fits naturally into this picture. Life too is a symmetry-breaking. In the beginning a homogeneous ocean somehow differentiated itself into cells and animalcules, predators and prey. Later on, a homogeneous population of apes differentiated itself into languages and cultures, arts and sciences and religions. Every time a symmetry is broken, new levels of diversity and creativity become possible. It may be that the nature of our universe and the nature of life are such that this process of diversification will have no end.

If this view of the universe as a steady progression of symmetry-breakings is valid, then Athens and Manchester fit in a natural way into the picture. The science of Athens, the science of Einstein, tries to find the underlying unifying principles of the universe by looking for hidden symmetries. Einstein's general relativity showed for the first time the enormous power of mathematical symmetry as a tool of discovery. Now we have reasons to believe that the symmetry of the universe becomes explicit and the laws of its behavior become unified if we go back far enough into the past. Particle physics is at the moment at the threshold of a big new step in this direction with the construction of grand unified models of the strong and weak interactions. The details of the grand unified models are worked out by studying the dynamics and composition of the universe as it is presumed to have existed for an unimaginably small fraction of a second after its beginning. Whether or not the particular models now proposed turn out to be correct, there is no doubt that the concept of

50

unifying physics by going back to a simpler and more symmetric past is a fruitful one. The science of Athens, what Einstein called the ancient dream that pure thought can grasp reality, is then nothing else than the exploration of our remotest past, "La Recherche du Temps Perdu" on a bolder scale than Proust ever imagined.

In a similar fashion, the science of Manchester and of Rutherford, the science of the diversifiers, is an exploration of the universe oriented toward the future. The further we go into the future, the more diversity of natural structures we shall discover, and the more diversity of technological artifice we shall create. It is then easy to understand why we have two kinds of scientists, the unifiers looking inward and backward into the past, the diversifiers looking outward and forward into the future. Unifiers are people whose driving passion is to find general principles which will explain everything. They are happy if they can leave the universe looking a little simpler than they found it. Diversifiers are people whose passion is to explore details. They are in love with the heterogeneity of nature and they agree with the saying "Le Bon Dieu Aime les Détails." They are happy if they leave the universe a little more complicated than they found it.

Now it is generally true that the very greatest scientists in each discipline are unifiers. This is especially true in physics. Newton and Einstein were supreme as unifiers. The great triumphs of physics have been triumphs of unification. We almost take it for granted that the road of progress in physics will be a wider and wider unification bringing more and more phenomena within the scope of a few fundamental principles. Einstein was so confident of the correctness of this road of unification that at the end of his life he took almost no interest in the experimental discoveries which were then beginning to make the world of physics look more complicated. It is difficult to find among physicists any serious voices in opposition to unification. One such voice is that of Emil Wiechert, whose contributions to physics were celebrated in a recent historical article by the physicist Res Jost, [Jost, 1980]. Speaking in Königsberg in 1896, the year after the discovery of X-rays by Röntgen, Wiechert said [Wiechert, 1896]:

"The matter which we suppose to be the main constituent of the universe is built out of small self-contained building-blocks, the chemical atoms. It cannot be repeated too often that the word "atom" is nowadays detached from any of the old philosophical speculations: we know precisely that the atoms with which we are dealing are in no sense the simplest conceivable components of the universe. On the contrary, a number of phenomena, especially in the area of spectroscopy, lead to the conclusion that atoms are very complicated structures. So far as modern science is concerned, we have to abandon completely the idea that by going into the realm of the small we shall reach the ultimate foundations of the universe. I believe we can abandon this idea without any regret. The universe is infinite in all directions, not only above us in the large but also below us in the small. If we start from our human scale of existence and explore the content of the universe further and further, we finally arrive, both in the large and in the small, at misty distances where first our senses and then even our concepts fail us."

These remarks of Wiechert show him to have been extraordinarily far-sighted. At the time when he was speaking, Boltzmann, Mach and Planck, the leading theoretical physicists of Germany, were still engaged in bitter arguments over the question of the real existence of atoms. Rutherford was then a young student newly arrived in England from New Zealand, only just started on the road that would lead him to the discovery of alpha-particles and atomic nuclei. More than half a century was to pass before the flowering of experimental particle physics, which would reveal a whole new world of strange objects and strange interactions hidden within structures that are themselves far smaller than atoms. Wiechert's

52

vision of an endless and infinitely diverse universe was quite out of tune with the visions of Planck and Einstein which were to dominate physics for the next fifty years. His words were ignored. Even his experimental discovery of the electron, made independently of J. J. Thompson and a little earlier, was ignored. Instead of nursing a grievance against his unappreciative colleagues, Wiechert had the wisdom to move away from the exploration of the very small. He turned his attention instead to the exploration of the interior of the Earth. He found his niche as one of the founding fathers of the science of geophysics.

In biology the roles are reversed. A very few of the greatest biologists are unifiers. Darwin was a unifier, consciously seeing himself as achieving for biology the unification which Newton had achieved for physics. Darwin succeeded in encompassing the entire organic world within his theory of evolution. But the organic world remains fundamentally and irremediably diverse. Diversity is the essence of life, and the essential achievement of Darwin's theory was to give intellectual coherence to that diversity. The working lives of ninety-nine out of a hundred biologists are spent in exploring the details of life's diversity, disentangling the complex behavior patterns of particular species or the marvelously intricate architecture of particular biochemical pathways. Biology is the natural domain of diversifiers as physics is the domain of unifiers. Unifiers like Darwin are as rare in biology as diversifiers like Wiechert are rare in physics. Darwin had no peer and no successor.

Or perhaps I should say, Darwin has only one successor and his name is Francis Crick. In saying this, I am not expressing a judgment of the greatness of Crick as compared with other contemporary biologists, and still less am I expressing a judgment of the importance of molecular biology as compared with sociobiology. I am merely saying, Crick is a unifier of biology in the style and tradition of Darwin.

Ninety-eight years after Darwin published his "Origin of Species," Crick propounded and named the Central Dogma of molecular biology, [Crick, 1957]:

"The Central Dogma. This states that once information has passed into protein it cannot get out again. In more detail, the transfer of information from nucleic acid to nucleic acid, or from nucleic acid to protein may be possible, but transfer from protein to protein, or from protein to nucleic acid is impossible."

The propounding of dogmas is an unusual activity for a biologist. Even Crick does not spend much of his time propounding dogmas. He spends most of his time studying the details of particular structures, just as Darwin spent much of his time studying details of the taxonomy of barnacles. But at heart Crick is a unifier, and in the end he will be remembered as the man who stated in simple words the unifying principle of biology for the twentieth century, as Darwin did for the nineteenth.

Let me now pull together the threads of my argument so far. I am saying that every science needs for its healthy growth a creative balance between unifiers and diversifiers. In the physics of the last hundred years, the unifiers have had things too much their own way. Diversifiers in physics, such as Wiechert in the 1890's and John Wheeler in our own time, have tended to be pushed out of the mainstream. John Wheeler holds uncompromisingly to his diversifier's view of the physical universe. Here is the theme-song of Wheeler's recent book, "Frontiers of Time," [Wheeler, 1978]:

"Individual events. Events beyond law. Events so numerous and so uncoordinated that, flaunting their freedom from formula, they yet fabricate firm form."

It sounds like Beowulf, but it is authentic Wheeler. This vision of nature is regarded by the orthodox physicists as belonging to poetry rather than to science. Wheeler's colleagues love him more than they listen to him. The physics of the unifiers has no room for his subversive thoughts.

In biology there has been a healthier balance. The mainstream of biology is the domain of the diversifiers, the domain of events numerous and uncoordinated, flaunting their freedom from formula. But when a unifier like Darwin or Crick arrives on the scene, he is not ignored. He is even, after a while, honored and rewarded. And his ideas flow into the mainstream. I am suggesting that there may come a time when physics will be willing to learn from biology as biology has been willing to learn from physics, a time when physics will accept the endless diversity of nature as one of its central themes, just as biology has accepted the unity of the genetic coding apparatus as one of its central dogmas.

III. Biology and Cosmology

The history of science is full of dichotomies. Athens and Manchester, past and future, unity and diversity, the universe of the cosmologist and the universe of the biologist. At various times in the historical development of science, one side or the other of these dichotomies has been over-emphasized. Sometimes unity and abstract structures are over-emphasized. Then the universe is seen as a solution of the equations of mathematical cosmology and our future is seen as determined by our past. Sometimes diversity and richness of detail are over-emphasized. Then the universe is seen as a habitat for real or imaginary animals, and our future is seen as open to the creative influences of random chance and free-will.

There was once a time when the ideals of unity and diversity in science were briefly held in balance. This was in the seventeenth century, when modern science was in its first flowering and both Descartes and Bacon were honored.

55

There was then no clear separation between the sciences of cosmology and biology. Even the most austere and respectable physicists conceived of the heavenly universe as filled with living creatures. Christiaan Huygens, originator of the wave theory of light and a physicist of impeccable credentials, wrote a book on cosmology with the title "Cosmotheoros," [Huygens, 1698]. "A man that is of Copernicus's opinion," he wrote, "that this Earth of ours is a Planet, carry'd round and enlighten'd by the Sun, like the rest of the Planets, cannot but sometimes think that it's not improbable that the rest of the Planets have their Dress and Furniture, and perhaps their Inhabitants too as well as this Earth of ours." His great contemporary Isaac Newton carried such thoughts even further, [Manuel, 1974]:

"As all regions below are replenished with living creatures, (not only the Earth with Beasts, and Sea with Fishes and the Air with Fowls and Insects, but also standing waters, vinegar, the bodies and blood of Animals and other juices with innumerable living creatures too small to be seen without the help of magnifying glasses) so may the heavens above be replenished with beings whose nature we do not understand. He that shall well consider the strange and wonderful nature of life and the frame of Animals, will think nothing beyond the possibility of nature, nothing too hard for the omnipotent power of God. And as the Planets remain in their orbs, so may any other bodies subsist at any distance from the earth, and much more may beings, who have a sufficient power of self motion, move whither they will, place themselves where they will, and continue in any regions of the heavens whatever, there to enjoy the society of one another, and by their messages or Angels to rule the earth and converse with the remotest regions. Thus may the whole heavens or any part thereof whatever be the habitation of the Blessed, and at the same time the earth be subject to their dominion. And to have thus the liberty and dominion of the whole heavens and the choice of the happiest places for abode seems a greater happiness then to be confined to any one place whatever."

I quote this passage from a recent book, "The Religion of Isaac Newton" by Frank Manuel. Huygens and Newton belonged to the last generation of cosmologists who felt free to people the universe with creatures in this fashion. The style and temper of science were already changing in a direction which would make such flights of fancy unacceptable. Neither Huygens nor Newton had the courage to expose their speculations to the ridicule of the public. Huygens arranged for his "Cosmotheoros" to be published after he was safely dead. Newton was even more timid and never published his private thoughts at all. But he was careful to preserve his manuscripts, and they can now after 300 years be found, still unpublished, in the library of the Hebrew University in Jerusalem. As Frank Manuel [Manuel, 1974] remarks: "This extensive text proves beyond question that Newton's worldview in the decade when the *Principia* was composed admitted of a far greater diversity of beings than those recognized by positivist physical scientists and nineteenth-century Unitarians." But Newton's manuscripts stayed hidden from his contemporaries in a big black box, and the eighteenth century knew nothing of them. The eighteenth century dawned bleakly under a heaven grown empty and dead. Cosmology, ever since that time, has concerned itself only with an empty and dead universe.

It is fruitless to speculate upon the might-have-beens of history. What might have been the effect on the cultural history of Europe in the eighteenth and nineteenth centuries, if Newton had had the courage to publish his speculative cosmology? Judging from the state of the manuscript, which is a finished copy accompanied by several rough drafts, he had at one time intended to publish it. If he had done so, placing his unparalleled prestige as the supreme intellect of the Age of Enlightenment behind an unashamedly romantic and poetic view of the cosmos, would it have made any difference? Would European culture have avoided that whole disastrous split between the narrow rationalism of the Enlightenment, dominated by Newton's public image, on the one hand, and the excessive irrationalism of the romantic reaction on the other? Could we have avoided the political manifestations of this split in Napoleonic centralism on the one hand and nationalistic frenzy on the other? Could we have avoided the conflict between science and religion, the conflict which soured the intellectual life of the nineteenth century and continues to impoverish Western culture even today? Obviously it is unfair to hold Newton responsible for all these later misfortunes. It was not his fault that the universe beyond the earth turned out to be dead and empty. When Newton decided to suppress his youthful visions of the cosmos, he was only doing what every good scientist is supposed to do, abandoning without mercy a beautiful theory which turned out to be unsupported by experimental facts.

So we have been left since Newton's time with a cosmology in which living creatures play no part. Only a few heretics like Emil Wiechert and John Wheeler dare to express the view that the structure of the universe may not be unambiguously reducible to a problem in physics. Only a few romantics like me continue to hope that one day the links between biology and cosmology may be restored.

What can we do now to start building bridges between biology and cosmology? There are at least two things we can do. The first thing is to look very hard at the universe and search for evidence of life and intelligence in remote places. If we are lucky, we may find that Newton gave up too easily his universe of celestial beings, that the cosmos is not really so dead and empty as it looks. The search for evidence of extraterrestrial intelligence is a continuing enterprise in which many respectable astronomers are intermittently engaged. Francis Crick has from the beginning taken this enterprise seriously and supported the efforts of the astronomers with his advice and encouragement. He is perhaps a little more optimistic than I am in his estimates of the chances of success. But the job of searching the universe for traces of life is a job for observers, not for theorists. There is little that theorists can do to help, except to serve as fund-raisers and cheerleaders.

But there is a second way of building bridges between biology and cosmology, a way open to theorists rather than to observers. The second way is open, whether or not the universe turns out to be peopled with celestial friends and colleagues. The second way is to build general theories of the potentialities of life in the universe. Francis Crick has been active here too. A few days ago I received a letter from him saying "I am still interested in the idea of Directed Panspermia. Our slogan was 'Bugs can go further.' " By this he means that there is every reason to expect spores to play an essential role in the propagation of life in the universe just as they do in the propagation of life on earth. Spores are the natural way to package biological and genetic information for rapid transit over interstellar distances. Panspermia is an old theory, originally proposed by the chemist Svante Arrhenius, a contemporary of Emil Wiechert. Arrhenius imagined the whole universe filled with the spores of life. Directed panspermia is panspermia plus intelligence, the universe filled with spores deliberately aimed toward habitats favorable to life's spread and survival, [Crick and Orgel, 1973].

Directed panspermia is only a hypothesis on the wilder fringe of speculation. It is not quite science and not quite science-fiction. It belongs with Newton's celestial zoo in the borderland where science and mythology meet. It is characteristic of future-oriented science that it easily runs astray into undisciplined speculation and outright fiction. Newton's timid soul, afraid to let his imagination run free, retreated to the safe ground of conventional science and conventional theology. Fortunately, Crick is a braver man than Newton, and our age is more tolerant than Newton's of scientific heretics. The prospects are bright for a future-oriented science, developing upon the foundations which Crick and others have laid, joining together in a disciplined fashion the resources of biology and cosmology. When this new science has grown mature enough to differentiate itself clearly from the surrounding farrago of myth and fiction, it might call itself Cosmic Ecology, the science of life in interaction with the cosmos as a whole. Cosmic ecology would look to the future rather than to the past for its subject-matter, and would admit life and intelligence on an equal footing with general relativity as factors influencing the evolution of the universe. Such a science would be carrying forward the tradition of the Manchester Literary and Philosophical Society, the tradition of plain people turning to science as a way of improvement of themselves and their neighbors. As Disraeli said a hundred and thirty-six years ago [Disraeli, 1844]:

"It is the philosopher alone who can conceive the grandeur of Manchester, and the immensity of its future. There are yet great truths to tell, if we had either the courage to announce or the temper to receive them."

REFERENCES

E. N. Da C. Andrade, 1961. "Rutherford at Manchester. 1913–14," pp. 27–42 of "Rutherford at Manchester," ed. J. B. Birks, (London, Heywood and Co., 1962).

P. M. S. Blackett, 1954. "Memories of Rutherford," reprinted as pp. 102–113 in "Rutherford at Manchester," op. cit.

Niels Bohr, 1961. "Reminiscences of the Founder of Nuclear Science and of some Developments Based on his work," pp. 114–167 of "Rutherford at Manchester," op. cit.

S. Chandrasekhar, 1979. "Einstein and General Relativity:

Historical Perspectives," *American Journal of Physics, 47,* 212–217.

Francis Crick, 1957. "On Protein Synthesis," *Symp. Soc. Exp. Biol. 12,* 138–163.

Francis Crick and Leslie Orgel, 1973. "Directed Panspermia," *Icarus, 19,* 341.

Benjamin Disraeli, 1844. "Coningsby, or the New Generation." The quotation is taken from page 152 of the 1849 edition, (London; Longmans, Green and Co.).

A. Einstein, 1915. "Zur Algemeinen Relativitätstheorie," *Sitz. Preuss. Akad. Wiss.* 778–786.

A. Einstein, 1933. "On the Method of Theoretical Physics," (Oxford, Clarendon Press), translated in "Ideas and Opinions," (New York, Bonanza, 1954).

Thomas Henry, 1785. "On the Advantages of Literature and Philosophy in General, and Especially on the Consistency of Literary and Philosophical, with Commercial Pursuits," *Manchester Memoirs, 1,* 7–28.

Christiaan Huygens, 1698. "The Celestial Worlds Discover'd: or Conjectures Concerning the Inhabitants, Plants, and Productions of the Worlds in the Planets," London. See also Marjorie Nicolson, "Voyages to the Moon," (New York, MacMillan, 1948), pp. 60–62.

Res Jost, 1980. "Boltzmann and Planck: Die Krise des Atomismus um die Jahrhundertwende und ihre Überwindung durch Einstein," Preprint, ETH, Zürich.

Frank Manuel, 1974. "The Religion of Isaac Newton," (Oxford University Press), pp. 99–102.

Joseph Priestley, 1774. "Experiments and Observations on Different Kinds of Air," (London), p. xiv.

Arnold Thackray, 1974. "Natural Knowledge in Cultural Context: The Manchester Model," *American Historical Review, 79,* 672–709.

John Wheeler, 1978. "Frontiers of Time," Preprint, Center

for Theoretical Physics, University of Texas, Austin, Texas, p. 13.

Emil Wiechert, 1896. "Die Theorie der Elektrodynamik und die Röntgensche Entdeckung," *Schriften der Physikalisch-Ökonomischen Gesellschaft zu Königberg in Preussen, 37,* 1–48.

Form and Aesthetics in
Twentieth Century Music

Gunther Schuller

That honor* that I've just received makes me even more apprehensive than I would normally be in such august company, as I find myself on this occasion; and I must confess that it fills me with some awe and considerable trepidation to find myself here in the midst of a community of scientists, scheduled to speak in a conference dedicated to the theme of "The Aesthetic Dimension of Science." This is not merely because we non-scientists tend to hold the scientific fraternity in such exalted veneration, not merely because scientists enjoy such an enviable status in our society unattainable by artists, but rather more that it is a deeply engrained belief, right or wrong, that the work and product of scientists flow from the domain of logic and thus constitute an "exact science," while the endeavors and creations of artists, musicians for example, comprise an inexact science, if a science at all. In a world in which the fruits of science and technology are

* Professor Schuller received the Gustavus Adolphus College Fine Arts Award for his distinguished contribution to the arts.

considered more useful—because more readily measurable than those of the artist—we musicians find ourselves on a somewhat unequal footing.

To be sure, the prevailing wisdom that scientific inquiry and discovery are inherently more capable of objectivity, while artistic creation is essentially a subjective, unquantifiable process, may upon closer inspection turn out to be myth and illusion. But there is little in our educative and societal environment that counters or disputes this myth. And so we artists are perceived as operating in a world which in no way relates to that of science and technology.

I think the planners of this conference think otherwise, and indeed hope to discover in our discussions during these days commonalities, analogies, parallels between and among these fields that need to be better articulated and understood. Feeling much the same way, I find that my qualms are mixed with a deep sense of privilege at being here, with the hope that these discussions can help to illuminate some of the mysteries that not only surround these respective fields, but usually keep them segregated and isolated from each other.

Of course, millions of words have been written and spoken over the centuries, even millennia, about the relationship of art and science. Let me just enumerate a few examples. We all know, I'm sure, about the various theories which equate, for instance, music and mathematics via the common bridge of acoustics; or the long-held belief that music is but a branch of mathematics; or for that matter the consistent pairing of the terms "science" and "art" from Greek antiquity on to the present day; or the ancient theories, still widely held, that music is but an audible exteriorization of the vibrations of life forces themselves—"vibrations of the cosmos" and "the music of the spheres" are the catch phrases of two of these notions; or again, the countless examples in the music literature of various mathematical or numerological approaches to the art of musical composition, from Johann Sebastian Bach to Alban Berg, John Cage and Milton Babbitt. Or the various concepts of constructivism in music, especially that of symmetrical constructions observable in the works of a number of early twentieth century composers like Schönberg, Scriabin, Bartok and Charles Ives; and in our own time the application, for example, of stochastic principles and those of information theory in, let us say, the works of composers like Yannis Xenakis, who is also a sometime mathematician and architect. Or the now-prevalent reality of computer-generated music where, obviously, twentieth century scientific and technological advances play a central role in the very creation and reproduction of music. And so on the list goes, with other lesser and greater instances of close ties between music and science.

Often enough such ideas were regarded, at least by their authors if not by the rest of the artistic community, as major breakthroughs, as innovations that purported to be panaceas which would solve in some objectifiable way problems which had previously defied resolution. One thinks of the highly-touted Schillinger system of composition and compositional analysis some decades ago; of the theories of the technologically oriented futurist movement in Italy and Switzerland around the time of World War I; or, I suppose, even the concept of Schönberg's "method of composing with twelve tones which are only related to each other"—insofar as this concept was understood (but I think more often misunderstood) as a *system*, a mathematical system at that.

No, there hasn't been exactly a shortage of attempts to systematize music, to formularize it, to quantify it, to absolutize it, to objectify and rationalize it, almost always out of an urge to emulate science, to make out of the "science of music," as so many music theoreticians of the Renaissance loved to call it, a more exact and measurable science. And yet in nearly four thousand years of contemplation of the phenomenon we call music, from Ptolemy and Plato to Leonard Meyer and Schenker to this very day, no universally accepted or even universally understood definition of music, of its essence, of its aesthetics has been realized.

So if we speak of form and aesthetics in music and especially in twentieth century music, as the announced title of my address suggests that I do, we really stand before a formidable task where even (as is relatively easier in the sciences) a list, an accumulation of historical data and a list of achievements—not just claimed achievements, but real achieved achievements such as scientists can point to—eludes the musician and other artists.

But that's not altogether a bad thing because I think therein lies the extraordinary power and beauty of music. That precisely because it is as an art essentially non-utilitarian, at least in our Western civilization's concept of music and art, and because it can say nothing specific or incontrovertible, it therefore can say and be everything. Precisely because the purpose and essence of music defy unequivocal, scientifically demonstrable defining, music speaks to us in ways that are at once profoundly moving and deeply personal as well as infinitely variable and diverse.

So what we are left with and have always been left with are viewpoints, theories, opinions, intellectual or aesthetic positions, which are determined in turn by a multitude of factors, conditioned in lesser or greater degrees and in a staggering variety of ways by background, education, fashions, individual capacities, be they physical—in music, for example, the oral or auditory capacities—or be they intellectual; and for all we know now conditioned—we don't know this for sure—by genetic factors, and finally by aesthetic and human values not directly related to music extant in the society at large. And mixed into all of this—as if it weren't already complicated enough—is something we in music dare to call "progress." For we believe, or at least have been absolutely certain until quite recently, that the developments in Western music since, let us say, the twelfth century—comprise a single more or less steady arch of progress.

I think it's fair to say that even scientists have had to learn that not everything that has happened in science since the Age of Enlightenment can be said to describe "progress." We musicians—though some would disagree with me—have, I think, an even greater problem in measuring, let alone claiming progress.

Indeed for me progress, in the sense that we usually understand that term, is either unmeasurable in music or irrelevant, or both. In my view we are left only with the knowledge (which by the way is quite sufficient for me) that each era, each epoch, each period, and beyond that different regions and countries and nationalities, have their own aesthetics and styles which may, in some final day of judgment, be deemed to be all equivalent—equivalent at least in their potential, or in their realization at the highest levels of genius and creativity.

But we have no means even of objectively verifying that statement or conjecture that I've just made. For example, at first glance one might assume, and I'm sure most of us in this room would assume, that Beethoven and his aesthetics would rank some universal priority among musical forms of expression. After all his music has moved human beings for 180 years and continues to do so unabated to this day and presumably will do so in the foreseeable future. We who belong more or less to that same culture that spawned Beethoven like to think that his music contains some profound truth and communicative ability that transcends all people of whatever rank, class or education or race. We ascribe to it universality. And above all, we torment ourselves today with the notion that somehow Beethoven was on to something that we in the twentieth century, or for that matter musicians in the fifteenth century, cannot today and could not then achieve, right? No, wrong! For even that widely held belief of Beethoven's uni-

67

versality is not at all true. It is once again conditioned by all kinds of inconclusive evidence and debatable assumptions. Upon closer inspection it turns out that while *some* non-Western cultures, such as Japan's, have accepted Beethoven with open arms, just as they have obviously accepted Western technology and now excel in it, other cultures such as the Javanese or Indian or Arabic, cannot relate to Beethoven's music at all, even when all conditions for such acceptance appear to be propitious. All we can really say with certainty is that, *for the moment* and probably for some foreseeable amount of time, Beethoven's music seems to have a deep appeal, both potential and actual, for a wide segment of the population in *certain* cultures and *certain* human societies. But that's as far as one can go. One can neither prove nor disprove that Beethoven's popularity is universal, is permanent and invariant.

Music historians, musicologists and well-read musicians know that our history is filled with every kind of mistaken verdict which in hindsight then often seems incomprehensible to us, be it an example of premature acclaim or belated recognition, of under-evaluation or over-glorification, of acceptance in some places or by some people and not by other places and other people, and so on in endless variations. Thus we know—those of us who think about these things—that it is difficult even to assess the past. And therefore, how much more difficult—as I am being asked to do—to assess the present!

For even a cursory historical glance at our music-historical past reveals that not only have musical aesthetics defied unanimous interpretation or definition but, more crucial, the very essence and role of music itself has resisted complete, unequivocal defining, despite countless attempts for centuries and indeed millennia to do so. The most that can be claimed is that at *certain* times *certain* aspects of the total phenomenon of music have been *temporarily* defined. The earliest attempts go back to Alexandrian and Greek antiquity, to Ptolemy, Plato, Aristotle and one Aristides Quintillianos, who all subscribed at least to the concept of dividing music into two equivalent complementary spheres which they called *scientia* and *ars:* the science of music and the art or practice of music. This very broad and generalized initial definition was rearticulated in the Middle Ages by such music theoreticians as Saint Augustine, Guido of Arezzo and Boethius, who amended the earlier concept of music to include the notion that two additional determinants were essential: those of *tone* (as an acoustical phenomenon) and *number* (as a mathematical element). Or to formulate it in another way, the two prerequisite factors by that definition, were *sensus* and ratio: the senses and the mind.

But it was in the final centuries of the Middle Ages, just before the Renaissance, that two further and, this time, opposing interpretations became current. It was the beginning of the split between the sacred and the secular, where the former sought to impose morality and truth as interpreted, of course, from a theological point of view, while the other would be satisfied with theoretical knowledge and practical ability. These restricted and essentially opposing viewpoints, though maintained in the Age of Humanism and the Renaissance, were nevertheless somewhat reinterpreted and broadened to emphasize the *practice* of music, particularly that of the vocal art as defined by ability and skill.

But it was not until the Age of Enlightenment that the ancient definitions and concepts of music finally gave way to an interpretation in which the *ratio,* the rational, the mind embodied in man—as opposed to God and religion and as delimited by *man's* intelligence—reigned supreme. And what is fascinating to note here is on the one hand once again the preeminence of mathematics in musical theory practice, and on the other hand a brand new ingredient at that time: the intelligence, the intellect of man, *Der Geist des Menschens* (as

69

the Germans would have it), as particularly defined in the eighteenth century formulation of man's mind as the *subject* and music as the *object*. And thus the Cartesian notion of "I am what I know" came to the forefront in music.

Man was now perceived as standing in a central position in determining both the science and the art of music. This perception in turn allowed for yet another idea to surface, namely the demand that music pursue the ideals of consonance and beauty. As Jean Jacques Rousseau put it in typically French fashion—this is a paraphrase translation: "Music is the art of combining sounds in a manner which is agreeable to the ear."

It is at this historical point—the last quarter of the eighteenth century—that a dramatic shift occurred, an almost 180 degree turn as it were, ushering in the Age of Romanticism. Here *ratio* and the mind were supplanted by the expression of emotions and feelings as experienced by the individual, by the Romantic hero personified by a Byron or a Beethoven. While such leading musicians as the composers Carl Maria von Weber, Richard Wagner and the famous German music historian, Johann Forkel, fully subscribed to that "romantic hero" point of view, they also tried to elevate music to the realm of the metaphysical and the transcendental. Of course, we all know that Wagner had his opponents, not so much in Brahms himself, but in Brahms' champion, the famous Viennese critic Hanslick, whose aesthetic was embodied in the motto "Music is nothing but sound-enlivened form," (in German "tönend bewegte Form"), a reiteration of music as an absolute, abstract, *sui generis* expression. That split, exemplified in the music of Wagner and Brahms, was perpetuated through the first half of our twentieth century by those two latter-day giants of music, Schönberg and Stravinsky.

Curiously Schönberg, although caught up as a young man in the Wagnerian and post-Wagnerian fever, nevertheless was an equally ardent admirer of Brahms and his brand of classic formalism. One can say that not only were the conflicts in Schönberg's music between the aesthetics of Brahms and Wagner never fully resolved, but in a fascinating alchemical fusion and synthesis, Schönberg managed to draw fruitfully upon both philosophies at various periods in his life. Still, when Schönberg defined his aesthetic in words and not in music, he combined his basically Romantic posture with one of the fundamental concepts of antiquity, namely music as an imitative art—and by that I don't mean mere program music; in other words an art not merely imitating nature's externals, but its *inner* expression through emotional and spiritual states.

Stravinsky, on the other hand, negated such a romantic viewpoint in favor of the old Middle Ages concept of music as pure expression of cosmic order. He said—and oddly enough his music often belies it—that music is incapable of expressing *anything*, even feeling, emotions, states of mind, let alone objects—which by the way, earlier in the century in a ludicrously boastful moment, Richard Strauss had arrogantly claimed for himself and music. He once said he could describe even a fork and a knife in music. Stravinsky felt that, what we call the "expressive" in music, was but an illusion. Whether he maintained this belief even after he converted at the end of his life to Schönberg's and Anton Webern's serial technique and its aesthetic is, oddly enough, not known. Stravinsky was to my knowledge never questioned on that subject and, as far as I know, he never voluntarily expressed himself on it in public.

Well, this brings us almost to the present in this brief capsule history of aesthetics and form in music. When I was a young composer in the 1940's, just beginning to discover myself aesthetically and artistically, it was made very clear to me and my generation that you absolutely *had* to choose between Schönberg and Stravinsky as leaders, not only between the atonal serial technique of the one and the tonal neo-classi-

71

cism of the other, but between their then very hotly debated philosophic and aesthetic positions. And I believe I was one of the first of my generation to say in response to that demand, "Nonsense. They are both great masters. I have much to learn from both."

Now, some forty years later, what seemed like such a complex and emotional choice pales by comparison with the present situation, where not two alternatives vie with each other but dozens. Therefore it is particularly difficult right now to say anything definitive about the direction of music today, about form and aesthetics in this part of the twentieth century. Our century is now eighty years old and in these short 80 years, particularly since 1945, our century has witnessed an unprecedented proliferation and fragmentation of viewpoints, of musical aesthetics, of philosophies and techniques and schools of thought. Moreover, none has emerged in a central or leadership position. What we have instead is a broad spectrum of musical categories and concepts, all in a vigorous state of flux and cross fertilization. This is an entirely historically unprecedented situation, awesome in its complexity, but aglow with the excitement of freedom—freedom of choice and of alternatives.

We are in my view on the threshold of some very exciting prospects. The certainties—or I have to say the *alleged* certainties of the avant-garde of but 15 and 20 years ago—have given way to the relative uncertainties of the present state of eclectism and pluralism. We were so sure in the late 50's and 60's that the prevailing aesthetics and principles of form were not only superior but inviolate. Not that everyone subscribed to such dicta, as they were handed down by the avant-garde centers in Europe (and in our own country) and its would-be dictators. But in the professions, in the academies and even the marketplace, it was made pretty clear in which direction music was supposed to be going at that time. And the dominant aesthetic was that of absolutism, the belief in the unassailable priority of absolute music; and the dominant forms were those of what came to be called the 'open form' and asymmetry. Pierre Boulez and Karlheinz Stockhausen were absolutely convinced that the 'closed' forms of the past, those forms tied to the functional diatonicism of the previous 2½ centuries were exhausted, were dead, were obsolete, and the aesthetic ideals that went with those traditional forms, both the classic and romantic and early twentieth century models, were also all declared dead and irrelevant.

Well, now at least, we seem to know better. It seems to me that we are and have been moving for about ten years toward a juncture where we are finally assessing in depth the innovations and experiments of the past 70 to 80 years, and both synthesizing and sorting out, with an eye (and an ear!) towards discovering *what* of all this recent past is of lasting value for now and for the future; and what on the other hand is unnecessary ballast. It seems to me that we are in a situation analogous to the one in the last third of the eighteenth century, when the previous innovations and experiments and various attempts to form a common musical language, a *lingua franca,* innovations which began with Monteverdi and culminated with Bach and three or four of his most gifted sons, resulted in a setting in which Haydn and Mozart had only to gather up various strands and pull them together into new forms, a new aesthetic and a new musical language. Beethoven in turn was to profit immensely from their efforts. We have gone through a similar period in this century, of pushing the frontiers of musical techniques, of new forms and aesthetics—new ones it seemed every new generation, every new decade—to a point where we have exhausted our hunger for experimentation and are now hoping to reap the benefits of those hard-won skirmishes and battles.

As I have already implied in this brief retrospective of the history of aesthetics and music, not all of our achievements in recent decades can be described as gains. I for one am not sure at all that the espousal of the 'open form,' for example, first envisioned by Claude Debussy and the young Stravinsky and elevated to a veritable prescription and dictum by Boulez, constituted a gain over the closed forms of the past. It is more likely that the open form represents an interesting alternative or addition to our form arsenal and resources. The 'open form,' an exciting discovery early in the century, did away with recapitulation, with repetition, with the preeminence of a central unifying musical idea and, in short order, with the very concept and use of theme. Melody was soon to follow. In the first flush of excitement at what seemed like boundless vistas of freedom, i.e., freedom from previous form constraints, it was difficult to see that such a concept of highly individualized and totally open form had in it the seeds of its own eventual demise. For what now followed was the assumption by composers who regarded themselves as progressive, that henceforth *every* piece by *every* composer could have—and some said *should* have—its own indigenous form, unique and peculiar to itself.

It took some 60 years to come to the realization that that was an untenable idea. There are not tens and hundreds of thousands of new or different forms. One composer in a thousand is fortunate if he can in one lifetime invent *one* form which is special, which is unique, which is clear and strong, and which is worth remembering. And then there is the basic contradiction, that, if it is worth remembering and thus worth using again, then why persist in a theory that form must be forever self-renewing. It seems to me that such thinking—and this was only 20 and 30 years ago—represented a basic misunderstanding of the very meaning of form in music. In the old days form used to mean a structure, a *recognizable* structure, a basic mold into which a music could be poured, as it were. But form as viewed in the recent past in the concept of 'open form,' i.e., as something constantly variable, unpredictable and finally anarchical, wasn't form at all. It was merely a container, a loose wrapping; and, alas, very often an indiscriminate *absence* of form.

It seems doubtful to me that certain traditional musical forms—whether the 'closed' sonata or variation forms of classical music, or the Blues form and structure in jazz, or the 'call and response' forms of many ethnic and folk musics around the world, particularly those of Black Africans and Native Americans—that such forms (and many more like them) will ever become useless or exhausted. Certain generic strains, even in music, seem to have survival built into their genes.

It is the task of the musical artist to know which these are. Ironically, many people, musical laymen, seem to already know. With the more recent recognition (in the last decade) that many of the traditional precepts of form can still be valid for our time, came the further recognition that we now have available to us a whole range of forms, running the gamut from the most sprawling and informal *wide open* forms—if we can manage to control them—to the most formalized, strict *closed* forms; and, of course, as usual a great many shadings and gradations in between. And I think it is not precluded that, just as Josef Haydn took the embryonic symphonic form of his day and developed it into its first full flowering, so too today a Haydn of tomorrow may come up with a brilliant new form concept.

In the realm of aesthetics things are even more kaleidoscopic. We are looking today in music at a bewildering array—a giant rainbow—of aesthetic alternatives of every conceivable shade and coloring, from John Cage's aesthetics of 'non-art,' 'indeterminate form' and 'chance operations' applied to all elements of music, to the exquisite and subtle refinements of virtually total control of a Milton Babbitt. In his music, for the first time in centuries, form equals content and content equals form. With a thousand and one variations and deviations in between those two extreme ends of the spectrum, we have thus a veritable plethora of aesthetics to choose from. Again the dichotomy of apparent total freedom of choice and the awesome responsibility of how and what to choose.

So, the young composer of today doesn't have it easy. On the one hand he can do everything and anything, singly or in combination, as he wants. But on the other hand, what a frightening challenge that is. How much one must know to make an intelligent choice among so many alternatives in this veritable *cafeteria* of musical aesthetics and techniques. Add to this new media and instrumental techniques, such as computer and electronic music, and a whole new world, for example, of multiphonics on modern wind instruments—where you play two and three or four notes simultaneously—and you have an idea of both the richness and the inherent complexity (some would call it confusion or chaos) of the present situation.

One very healthy sign is that composers seem to be thinking about aesthetics and beauty again. Beauty was an inadmissible and bad word just 20 or 30 years ago. It was never discussed in a graduate class in composition. It was not long ago that the average young Turk, aided and abetted by the musical academicians that populate our schools and graduate music departments, equated aesthetics solely with technique. Such and such a technique begat such and such an aesthetic, and vice versa. It is true, of course, that certain techniques if used in a very restricted way, will result in similarly restricted aesthetic frameworks; and that again is precisely one of the choices a composer can still make, if he chooses to do so. But in contrast to yesteryear, the erstwhile clear lines of demarcation between formal and aesthetic alternatives—which in the past were often hotly contested, as I mentioned earlier—have now become blurred and softened; and formerly inimical aesthetic viewpoints have begun to converge, to overlap, to

76

cross-fertilize, or at the very least learned to coexist peacefully. Call that other approach 'musical isolationism,' 'musical aryanism,' or maintenance of pure genetic musical strains: we had all that and more for years. But now, in my view at least, we seem to enjoy a more benign, open-vistaed outlook.

Of course, I don't wish to be trapped into talking only about what we call 'classical contemporary music.' For it is precisely the present-day reality, in which all kinds of musics coexist on the face of this globe and, I dare say, now even cohabit in new and wonderful pairings and combinings—maybe even 'symmetry breakings'—which defines most accurately the present state of music, notwithstanding the fact that the entrenched musical establishments of our Western culture have not yet understood or accepted that reality. I regard the old notions of European-based 'classical' music as being inherently superior to all other forms and aesthetics of music as totally untenable. I have a quite different view of music; I call it the 'global view' of music; and in that global view there are only two kinds of music as Gioacchino Rossini put it a century and a half ago: "good and bad." There is no inherent or automatic correlation between *categories* or *types* of music and quality or the lack of it. There is only a broad spectrum of musics and musical aesthetics ranging from the most ancient to the newest, from the popular, folk, ethnic and vernacular to the serious, classical, symphonic—those are all hopeless misnomers—from the improvised to the fully notated: in all these different *kinds* of music there can be found the exalted creations of musical geniuses or the droppings of the mediocre and the feeble offerings of commercial parasites. No musical strain has an inherent claim on quality—or the absence of it. The genius of Beethoven and Bach or Louis Armstrong and Charlie Parker; the genius of those unknown musicians who eight or nine centuries ago created *Etenraku* and the other ceremonial masterpieces of Japanese *Gagaku;* or the genius of an African improvised drum ensemble,

77

whose rhythmic complexity and virtuosity defy our Western capacities of musical comprehension, let alone emulation; or the genius of the eloquent nature-bound ceremonial music and dances of our own Pueblo Indians; or the heavenly serenity of Javanese Gamelan traditions—I could go on citing many, many more—all these forms of musical expressions are in my view and to my ears *qualitatively* equivalent.

That is why the motto of my own publishing company reads "All musics are created equal." By the way: not all musics *are* equal, but all musics are *created* equal, that is to say, potentially equal in quality. It all depends on who chooses to express himself through these kinds and categories of music and *how*. Our Western habit of regarding all those other musics and aesthetics out there as "primitive"—how dare we call African drum ensembles primitive!—or irrelevant is today a totally untenable and, in my view, embarrassingly ludicrous idea. How dare we sit in judgment looking down from our European thrones on Javanese or African music or, for that matter, that of the Eskimos!

Well, one specific reason why we dare not, came to us through the courtesy of science and technology and the ingenuity of modern man: namely through the phonograph, the radio, television, satellite transmission and other twentieth century breakthroughs in communication. Before our time we did have some sort of an excuse for our ignorance and prejudices about other musics. We literally could not know empirically what we can know today. Beethoven had no way of learning anything about the music of Tunisia or Afghanistan or of the Japanese court. Perhaps a Marco Polo or a Magellan had some inkling of the world's musical diversity, but neither of those gentlemen let on if they did.

But today's budding musician or lay listener is in the most fantastic position; he can go to the nearest record store and buy not one recorded example of such music, the music of Afghanistani mountain shepherds or whatever it might be, but literally dozens and hundreds of such recorded examples. No, it doesn't behoove us—nor cover us with much glory—to continue to ignore that whole other beautiful universe of music that lies out there beyond our own narrow musical horizons.

And here we artists have, I think, a lesson to learn from science, the history of science, and the aesthetics which motivate most creative scientists. Perhaps not in all the sciences and certainly not until recent centuries, but it seems to me from my limited knowledge that science and scientists accept much more readily than artists do the infinite diversity of nature. Certainly in biology, since Darwin, and more recently in physics, let's say in the work of a John Wheeler, diversity is understood as the very essence of life. It puzzles me why we Western musicians and artists cannot understand and cherish that same diversity in our field. Why can we not appreciate and teach that music, especially as a creative expression, is not merely an entertainment, a commercial commodity, a salable product, but in its highest forms an expression or a reflection of *life itself,* in all its myriad and infinite diversity. In my very amateur but nevertheless hopefully correct view of the history and development of science, I see an extraordinary parallel to what I have called a 'global view of music:' namely if the development of the universe can be described as a never-ending breaking up—'symmetry breaking' as you call it—of cells and molecules and atoms and micro-organisms, then so too music has proliferated and fragmented in never-ending replication and differentiation from that first moment when primitive man uttered his first primeval sound. That is a concept of music that I find beautiful, for it is consonant with the development not only of life, but of humankind.

That seems to me to be a lesson which we musicians and artists in general have yet to learn from the scientists. The concept of 'unity and diversity' is, alas, foreign to most musicians—at the very time that modern technology has given us the tools and the technical capabilities of appreciating just that.

There is much more to be said about other matters that I haven't even mentioned, although some of these we touched upon briefly in some of the discussions yesterday. For example, the function and utility of music; or the role of beauty in music, of simplicity and of complexity and the interaction of the two; and above all the deep mysteries of the creative process, about which we know next to nothing. I have described to you where matters stand now in respect to concepts, schools, styles, philosophies of music and the extreme fragmentation that characterizes both the narrow field of Western classical music and the much larger field of global musical pluralism.

But there remains one other matter which I *must* mention, if only in passing, which is perhaps especially appropriate to discuss in the framework of this conference. It is a subject which I will have to state mostly in the form of questions, for it is not something to which I or anyone as yet has the answer. Moreover, it is a subject located at the very frontiers of both music and science and one which neither of us, musician and scientist, can study or solve independently, It is an area where we must collaborate and pool our talents and resources.

Formulated in the simplest terms the questions (or the problem) is: how precisely do we learn music? This may seem at first hearing a harmless or irrelevant or even stupid question. Many of you may say: of course, we know how we learn music. But the fact is we don't. The question's relevancy becomes agonizingly clear when we realize that there are many thousands of schools and tens of thousands of teachers who all teach music, without any of us really understanding *how* we learn and what the processes of acquiring musical skills, creative or recreative, are. In other words, how can we teach something effectively if we don't know how and by what

80

combination and accumulation of skills we learn what we learn?

I, for example, can tell you nothing about why I know what I know, and why I can do the particular musical things that I *can* do, and how I acquired those skills. And around this painful question cluster several other related ones. I cite only a few. What really are the formal dynamic principles that govern the creation of music—of this or any other period? What indeed is the *nature* of music? What is its essence, its content? What is the relationship between the human mind and human senses and the external world, which we presumably observe and, therefore, in partial ways understand and reflect? And why have we failed, thus far—as I pointed out earlier—to arrive at a complete formal and aesthetic theory of music? Ironically the answers to these questions elude us intellectually, even though any musician will tell you that such intellectual abstractions constitute the most concrete experiential realities of music, realities which when we experience them convince us positively that a life in music is *life itself,* is beautiful, the fullness of living. In other words we *feel* these things, but we can't explain them; we can't give them an intellectually cogent, coherent definition.

I don't know whether it is entirely fair to ascribe one of our problems in music (and the other arts) to science: namely, the dichotomy between objectivity and subjectivity. I suspect that this issue has been with us a long time, since antiquity in fact, although not always articulated as a problem. But since the advent of modern science the conflict between objectivity and subjectivity, between the absolute and the relative (or the conditional), between intellectual knowledge and direct experiences, instinctual sensory responses—all these dichotomies have become priority concerns for the music theorist and aesthetician. As mentioned earlier, there has been no lack of attempts to resolve these dichotomies and to unify objective and subjective values into a comprehensive and systematic whole. And yet, as I have also pointed out, our best minds have not yet achieved that goal of accurately defining the deeper formal and aesthetic realities of music. And because of this failure even the best of our music education, not only of professional musicians but of the general public, that is, our potential audience, is incomplete and inadequate unto the task. But how little we hear about these concerns and questions in the education offered by our music schools, conservatories and colleges and universities, virtually all careerist-oriented, where too frequently the acquisition of a minimum of utilitarian knowledge is considered sufficient to achieve a maximum of success and financial income!

I'm sure you have noticed by now that my knowledge of science is also hopelessly incomplete and inadequate. Nevertheless with my heart pounding, treading, I hope, not where only fools tread, I dare to suggest in closing an extraordinary and to me profoundly beautiful statement by Arthur Eddington. It reads: "All through the physical world runs that unknown content which must surely be the stuff of our consciousness. Here is a hint of aspects deep within the world of physics and yet unattainable by the methods of physics." Talk about the scientist as artist and philosopher and poet. Is this a hint, a possible postulate or hypothesis for further inquiry as to the true relationship of the objective, quantifiable world of *matter* and the subjective qualitative domain of the *mind?* For are not mind and matter ultimately derived from the same life sources? If I understand it correctly, quantum mechanics has shown that, at the basis of the physical universe, lies what—I hope I'm correct in this—quantum physicists have called the 'ground state' or the 'vacuum state'; and that this field and this conception is one not only of perfect order but of infinite energy and intelligence, whence all creation around us—so the theory suggests—has arisen. But there is another branch of science, psycho-physiology. If it is true, as the psycho-physiologists claim, that the thinking process of the human mind has at its basis what these scientists call the "state of least excitation of consciousness," then do we not have here a fascinating parallel between the physical universe and the human mind? That is, does not this suggest that these both derive from the same or single common source, ground source, *Urgrund*, as the Germans call it? And if *that* is the case, then do we not have here a postulate, a theory for re-solving those eternal conflicts that have plagued us forever and ever which I cited earlier: the dichotomies between ob-jectivity and subjectivity, absolute and relative, physics and metaphysics, the external and inner worlds, indeed, science and art?

Perhaps this is what the greatest minds from Plato and Aristotle to Einstein, Eddington and Wheeler have tried to get at or tell us; that the structure of the human mind and the external physical world are *at ground* identical or at least par-allel. My question is: can we musicians and scientists explore such an idea together, systematically and empirically, begin-ning perhaps with a conference such as this?

I am convinced we musicians need science's help. A more profound understanding of music depends on a more com-plete understanding of the workings and dynamics of the mind and perhaps then, working at the very frontier of con-sciousness where mind and matter meet, we will begin to understand the creative act, what we call "inspiration" and have not been able to fully describe or define; and how au-ditory perception and the acquisition of the skills related thereto really function. How close we artists and you scien-tists can be, and at the same time how deep the symbiotic relationship between man and his environment can be, is ex-emplified in this oft-quoted and most touching statement and revelation of Einstein, an appropriate thought, I think, to

83

leave with you as I conclude this talk: "The theory of relativity occurred to me by intuition, and music is the driving force behind this intuition. My parents had me study the violin from the time I was six. My new discovery is the result of musical perception."

Thank you very much.

Science as the Search for
the Hidden Beauty of the World

Charles Hartshorne

A scientist is one who wants to find out the truth by close observation. I derive this definition from the writings of Charles Peirce, who was a practicing scientist in several branches of knowledge, including physics, geodetics, and psychology. The definition suggests the query, "Why should one want to find out the truth?" Also the query, "Why must it be by observation that science arrives at the truth?" In regard to the latter question, Peirce held, I think correctly, that even in pure mathematics there is a kind of observation. This is inspection of what Peirce called diagrams in a special technical sense according to which the equations of algebra, or the symbolic arrays used in mathematical logic, are as truly, or even more truly, diagrams than those employed in plane geometry. Hence Peirce's definition covers even mathematics. So let us turn to the first question, why one should wish to find out the truth: Is knowledge of the truth, simply as such, desirable and if so why?

It is clear enough that there are truths that have almost

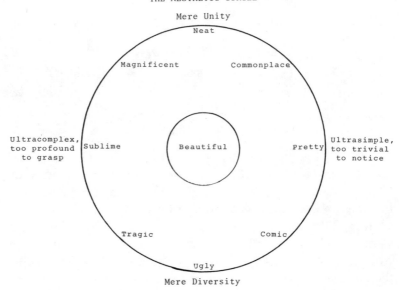

THE AESTHETIC CIRCLE

Mere Unity

Neat

Magnificent   Commonplace

Ultracomplex,   Sublime   Beautiful   Pretty   Ultrasimple,
too profound   too trivial
to grasp   to notice

Tragic   Comic

Ugly

Mere Diversity

(Outside the circle, no experience
and no aesthetic value.)

86

nothing to do with science. For example, the number or percentage of people in the phone book of a certain area whose first names begin with R and last names begin with P. It is hard to imagine any scientific relevance for this truth.

The truths sought by science are not of this kind. Of what kind then?

Before we try to characterize scientifically relevant truth, it may be well to remind ourselves that there are two ways in which science is valued, two basic reasons for wanting to be a scientist oneself or for thinking it well that there are scientists among us. One is that science, through its applications, is useful. The truths it discovers make technological progress possible, as everyone with a TV set, or an automobile, or access to a doctor and with symptoms of a disease is likely to realize. The transition from pure science to technological applications is made by engineers or inventors. Medical people might be called engineers of health. In our American tradition the term "scientist" is rather commonly used to cover these persons as well as the more purely theoretical inquirers and discoverers. There is some justification for so using the term; but as with other very broad usages of words, something is lost by this one. Maxwell, Einstein, Dirac, Heisenberg, invented nothing, I suppose, and made no application of their discoveries, yet without such theorists as these there would be no technology as we know it.

Engineers value truths for their applications, and good scientists rather expect there to be such applications. But they generally, almost unanimously, tell us that this is not the primary or most fundamental reason or motivation for their activity in search of truth. The most fundamental reason is the expectation that the truth of the sort they hope to find will be *beautiful*, to employ a word they commonly use in this connection. Mathematicians, astronomers, biologists, all speak of the beauty of their theories or formulae. A poet—and every feminist should rejoice that it was a woman—wrote of her experience of geometry as a student at Vassar College: "Euclid alone hath looked on beauty bare." Whoever does not feel something of this aesthetic aspect of mathematics can rest assured that being a mathematician would not have been a viable occupation for him or her, unless—as is likely enough—he or she suffers from having had poor teaching in the subject at school. I was in the audience when Dirac, the theoretical discoveror of anti-matter, said, "It is more important that a theory be beautiful than that it be true." He did not elaborate, as one might wish him to have done. But at least the statement, startling as it is, indicates that scientists do value beauty.

Suppose that, in some phone book, the percentage of persons having R for first initial and P for last initial is 5, or any other percentage you please, what beauty could one find in this fact all by itself? Also, what utility would the fact have? Isolated truths—to take another example, the average number of hairs in the heads of men fifty years old and with the middle name Henry—are ugly and useless to such an extent that no one tries to ascertain them. Life is far too short to bother.

Beauty and utility alike require more than mere truth. And this more is in good part much the same in both cases. A beautiful whole is at least an arrangement of parts felt or experienced as equally free from the opposite vices of monotony and chaos. Monotony is a lack of contrast or variety, chaos is a lack of similarity or unity. The ideal is unity in diversity, with equal emphasis on the two requirements. In the phone book example mentioned there seems no kind of unity or connection between the items taken together. The deficiency is in the direction of chaos, merely arbitrary togetherness. In a beautiful whole, experiencing one part or subset of parts prepares one for other parts or sets of parts. On the other hand, if this preparation goes too far, so that passage to the other parts provides no significant novelty, no element of surprise or important addition to the experience, then there is a sense of monotony, of idle repetition.

Granted that there is no excess of unity over diversity, of similarity over contrast, or vice versa, there may still be no profound beauty. A single harmonious musical chord has contrast and yet gives a sense of unity. But the beauty seems trifling, lacking in complexity or depth. It has contrast but not the higher order of contrast that obtains among a diversity of contrasts. For these one looks not to a single chord but to a symphony, or perhaps a song with both words and music. Beauty thus is a reasonably complex system of contrasts, an adequately diversified unity or adequately integrated diversity. Where the diversity, though proportional to the total complexity, is unimpressive, or on a superficial level, we may call the system pretty rather than beautiful, as in a simple flower. But a well-arranged flower garden is more than pretty, it can be beautiful. Where the diversity is very impressive indeed, to the point of almost overwhelming us with a feeling of our lack of ability to do justice to it, we may say the thing is sublime rather than merely beautiful.[1]

Contemporary physics is, in some ways, better unified than Newtonian physics, if both are taken as accounts of observable phenomena. And the diversity of observed facts it covers is incomparably greater. Evolutionary biology likewise incomparably surpasses all previous biology. To unify our ideas of the vast array of species by saying that God made them all, just as the Book of Genesis says, leaves the relationships among the species largely in the dark. I for one believe that God is, in a reasonable sense, universal Creator, but I fail to see in "special creation" of species any light on the how of the divine action or on the relationships among the phenomena. The idea is formless and unbeautiful in the extreme. It was Darwin who closed his first great book with a prose poem to the beauty of the web of life.

In addition to the word *beauty* as a mark of scientific truth there is also the word *simple*. But, if the value of beauty is proportional to the scope of contrasts unified, it is clear that the two criteria differ, at least in emphasis. It is even arguable that extreme simplicity is a mark of error rather than of truth. Aristotle's theory of genetics was far simpler in any reasonable sense than neo-Mendelian theory. And how simple is quantum theory?

Aesthetics has taken two thousand years to realize that the first noun in the formula "unity in variety" is no more important than the second. Greek and even medieval thought put the emphasis on unity. God was the absolute beauty and God was said to be absolutely "simple." The difference between God and any creature was also simple; the attribute of deity was infinity rather than finitude, or absoluteness rather than relativity, immutability rather than mutability, simplicity rather than complexity. No simpler intellectually respectable way to contrast divine and not divine has ever been proposed. What is called process theology, or the theory of "dual transcendence," holds that the real contrast is double, e.g., infinity contrasted to finitude, but also an eminent, uniquely excellent form of finitude contrasted to ordinary finitude. God is infinite in whatever sense infinitude makes sense *and* is an excellence, and God is finite in unrivalled fashion in whatever sense—and there is such a sense—finitude is an excellence and mere infinity is not. Similarly with absolute versus relative, mutable versus immutable, etc. This view is a good deal more complex than the traditional one. Why should the characterization of deity be so simple and easy as the mere distinction between finite and not finite or mutable and not mutable?

One of the ways in which simplicity is a possibly misleading criterion is seen in considering artistic beauty, say in music. It has become a commonplace in musicology that a completely orderly, and hence in principle completely predictable, succession of sounds is not beautiful. An element of unpredictability is required. If the *merely* unpredictable is

90

chaotic, the merely predictable is monotonous, tedious, lifeless. "Aesthetic order is the vast realm between the fatal extremes of mechanism and chaos." Kurt Sachs in this epigram sums up his penetrating insight into the phenomena.

In natural science, from Democritus to Einstein, the ideal of science was sheer determinism, absolute order, leaving nothing to chance or caprice. Theories of art for centuries tended to view artistic order in similar terms. Every part was supposed to imply every other. So, did the first word of a poem (which might be the first personal singular pronoun) imply the whole poem? The first two words? How ridiculous this was seemed to escape many. Artistic order is not, even in ideal, so deterministic. Is the correct scientific ideal deterministic? Epicurus did not think so, Aristotle did not; Peirce, Bergson, and Whitehead did not; and some of the founders of quantum physics did not. I hold that they were right.

Determinism is the simplest view of causality. Situations are seen as both necessary and sufficient conditions for what subsequently happens. All chance is excluded. The matter is controversial, but many authorities now hold that even Newtonian physics need not be interpreted as strictly deterministic, still less quantum physics. Events have necessary conditions, but "sufficient condition" is a limiting concept not to be taken as literally true of concrete happenings. The true laws are approximate or statistical. Human freedom in the sense of actions not casually determined is merely a special, high-level form of what Peirce called "spontaneity" and Whitehead and Bergson, "creativity" and held to be pervasive of nature.

Beauty in natural science is intellectual (Shelley wrote a poem to "intellectual beauty" and he had science in mind); beauty in art and much of life is in good part emotional and sensory. Let us consider this contrast. Intellectual search for truth itself supports the view that there is no such thing as intellect, if this means a self-sufficient function or activity,

91

and no such thing as disinterestedness, if that means an emotionless recording of fact or solving of problems, as cameras or computers record or solve. Psychological observations reinforce and particularize the old theological view that the motive of all healthy human life is love, which in diseased form is resentment or hatred. The only possible disinterestedness capable of producing knowledge must spring from a faith, and in part an experience, that reality when truly known is lovable. One may also say, as Poincaré and others have said: We seek the truth about nature because we feel that nature is beautiful. Intellectualist theories deny this or, for most purposes the same thing, reduce it to a dull platitude.

Are pragmatic theories much better? They are often so put as to imply that the basic problems are mere puzzles without any characteristic emotional content. To survive or maintain the species one may choose this means or that. To reach a certain end, the feasible means are such and such. It is assumed that the organism wants to survive, to retain its powers, that it has goals and wants to attain them. But the role of emotion is left obscure. In spite of Freud we still hug our illusions about the independence of the articulate aspects of experience.

What is science concretely? For the sake of emphasis, we may start with the extreme Baconian formula: observation plus explanatory hypothesis. Since "observation" is a psychological term and hypothesis a logical one, to make our ideas parallel we should rather say, observation and imagination. But then it becomes plain that the order needs to be reversed. Only infants and the lower animals (if even they) observe without prior imaginative interpretation. Non-scientific cultures are full of theories. Science began, as Popper so well says, with the "criticism of myth." There were plenty of "hypotheses" long before what we call science; and as for observation, one can see a dog or cat observing busily. The unique thing is to observe for the purpose of supporting a belief against criticism, or of supporting a criticism against the belief. In either case, one must be interested in the belief and in the criticism; one must somehow, as it were, love or hate them. No one now bothers to criticize the Homeric religious myths; they are neither loved nor hated, at least not as beliefs that might be true. Hence they inspire no scientific activity. But there are beliefs we do care about and which also some are moved to criticize; both parties may then appeal to observation because they find no other effective means of adjudication.

Why do we imagine, or form ideas going beyond the obvious deliverances of perception? Pragmatists have a ready answer: it is more efficient to deal with the perceived world in terms of the remembered, suitably transformed in imagination to fit the partly new situation, than merely to respond instinctively to the perceived. Animals with imagination can deal both with the repetitive and the unprecedented; those without it cannot. Natural selection will therefore favor animals with suitable amounts of imagination. However, natural selection does not of itself or directly provide motivation for behavior. Modes of behavior, not feelings, are directly selected for; however, animals do not deliberately imagine because they know that this will help the species to endure and expand. Rather, animals must be able to enjoy imagining; and then natural selection may favor those which have unusual amounts of this capacity and hence behave in unusually flexible ways. There must be an inner reward of imagination, distinct from its eventual utility. Any child illustrates this reward. Imagining may produce intensity and harmony of experience, which in the broad meaning is aesthetic value or beauty. (In insanity this inner reward of imagining is so much greater than that of the recognition of perceived reality that the latter is renounced in order to enjoy the former without distraction.) The beauty of hypotheses, we may say, is the psychological ground of their possibility. Mathematicians, physicists, biologists, often say this; logicians of science are somewhat grudging in their admissions of it.

The particular form taken by our hypotheses is subject to at least three influences: first, our sensitivity to the harmony and variety which they contribute to our imaginative experience; second, the special slant of the individual's emotional and imaginative habits; third, the extent to which discord with perceived reality is avoided. Apart from science, and the discipline of daily life, this last influence is often weak: a pretty picture, if unconnected with daily practice, may be accepted in spite of fairly obvious discrepancies with fact. Medieval bestiaries illustrate this. Who that really cared about truth concerning animals would have believed some of the grotesqueries solemnly proclaimed in these texts! (True, the animals were mostly attributed to distant, terrifyingly unknown lands.) Science begins when observational testing of imagination becomes a serious occupation. This happens when people begin to want the truth about something, when conflicting imaginings lead to the search for a neutral arbiter, and when authority can no longer provide this because of the multiplicity of pretended authorities. Conflict of imaginings is felt to be unsatisfactory because of the social dimension of experience. One cares enough about the other fellow to internalize his imaginings, and is then disturbed by their lack of accord with one's own. The critical attitude is essentially social. Self-criticism is chiefly a product of criticism by others, and could not long flourish without it.

What we have then is this: imagination, though practically useful, is not indulged in directly for its utility but for its intrinsic value, though in the long run its utility may determine its evolutionary development. It is observationally tested when it is undecided between alternatives or when sympathy for the views of others causes resort to the only available arbiter, perception. But why do we not simply choose between our own view and the critic's on the basis of the relative attractiveness of the two conceptions? The answer can only be that we care enough about reality itself, the perceived cosmos, to feel that an inner harmony of imagination which is also in accord with the universe will be more rewarding than a merely internal harmony alone. To pragmatists it may seem clear that this extra reward can only be due to practical utility. However, the great scientists do not interpret themselves in this fashion. Their faith is that nature is more beautiful than our self-indulged imaginative creations can be. And looking back, one sees a confirmation of this faith. The real world keeps turning out to be more thrillingly beautiful, in a vastly greater variety and intricacy of ways, than our previous dreams. How dull and unaesthetic the Newtonian picture seems now; how messy the notion of special creation in biology, or of the movements of heavenly bodies in Ptolemaic astronomy; how tame the tight little universe, and its four thousand years of development, which was all that the Middle Ages in Europe contemplated by way of the "material sublime." One could go on and on. The romance of science is its disclosure of a universe whose wild harmonies surpass the most vivid dreams of imagination not submitting itself to criticism and observational test.

Why have European physicists, until recently, provided nearly all the basic ideas, while Americans (always excepting Willard Gibbs) have furnished chiefly the ingenious and careful experiments? Can it be because in this country sober unimaginative practicality for long overbalanced the taste for the romantic beauty of the cosmos? I see no other reason, unless it be that the waste of bright pupils' time in our mass-educational schools produces a sense of inadequacy which inhibits speculative daring. Take your choice. Probably both factors have operated.

Science we may now define as essentially a form of love or sympathy, sympathy for the ideas of others and love of reality as open to observational inquiry. It is the imaginative, socially critical, and observational feeling for nature. One may also say, it is love of nature in depth, rather than merely for its obvious surface beauties, or (as in some poetry) its fancied connections with hidden things in no way even indirectly disclosed by perception. It expresses the faith that mere common sense and mere dreams are both dull or chaotic things compared to the discoverable riches of the actual world. Science is a great romance, in a way the greatest.

So far we have spoken of science as though it were but one thing, with one principle and one method. But the beauty of truth lies in part in its diversity and radical contrasts. Any mere leveling of differences between sciences, even as to method, is anti-scientific. It kills the romance of the quest for truth by what at best is uninspired and uninspiring vagueness.

So let us consider now the kinds of science, or the imaginative, observational, critical love of nature—or reality. An inquirer inspired by this love may proceed in either of two chief directions: one may seek to detect general laws in particular or individual cases; or one may seek to grasp what is peculiar about an individual case, applying to it such laws as are known. Physics illustrates the former, human history and natural history the latter. Both are finally united in cosmology, the natural history of our universe or cosmic epoch. In its totality, as Croce and Neurath held, knowledge of the contingent is historical rather than systematical, although it includes a systematic element. This illustrates the principle that becoming is the ultimate category, not being. For science, at least, this seems to be so. The explanation of a contingent thing is its genesis. Even laws (the most general ones) if contingent must be explained, if at all, by their genesis. The alternative is to declare them inexplicable. To pursue this theme would very likely take us beyond empirical science to metaphysics or theology.

That natural science is finally historical implies that the

social sciences cannot be divided from the natural merely on the ground that while humanity has a history, nature in general does not. Darwin and the great geologists and astronomers lived in vain if we accept this. But there is another apparent ground of division. Natural science, it seems, deals with matter; social science, with mind. There are then three possibilities for the long-run development of science: we shall end up with a physicalistic monism; or with a psycho-physical dualism; or with a psychicalistic monism. It is evasion to talk avout "neutral monism," "evolutionary naturalism," or "emergence." The emergence of mind as such simply means that physicalistic monism was once the whole truth about nature, but that dualism is now that truth. As for neutralism, there is no science of the neutral. Matter is studied, and some psychologists at least think that they study mind; but who is practicing imaginative, critical, observational study, inspired by a love of neutral stuff? I know of no such person.

A mere dualism is unscientific, since it is a denial of the unity upon which all beauty depends. And science is the sense of the discoverable beauty of things. But shall the unity be in physical or in psychical terms? In *method* science must be physicalistic and behavioristic, since only behavior is inter-subjectively observable. But "method," "intersubjective," "observable," are psychical terms; and so behaviorism taken as a materialistic monism is inconsistent with its own argument. Hence it is not surprising that the case for behaviorism was first sharply stated by Peirce, a psychicalist. And indeed only psychicalism can consistently practice behaviorism while avoiding dualism. Its view is that matter, including behavior, is how mind on certain levels is given or appears to other mind; it simply is mind as spatio-temporally structured and interrelated. Psychical terms are indispensable; it has been an open secret since Leibniz that "matter," taken as wholly other than mind, is dispensable. That minds experience other minds is the social character of mind itself; and the spatial extendedness of a community of minds is deducible from this social character. No additional concept of matter is needed, nor does it add anything to the conceptual system already set up in terms of mind.

All natural science can then be unified in a cosmic natural history (in principle, perhaps, inclusive of the history of laws, though this is problematic), a natural history of various kinds of minds, whose lower levels constitute what is called matter. For the sake of intersubjectivity, this system of minds must be dealt with behavioristically; but philosophically, we should take behavior as simply the way in which various interrelated levels of mind are observable by minds on our level. This view is romantic enough to fit the romantic nature of science itself. Sober persons who have never quite caught the romance of inquiry are shocked by this feature. I think it is a recommendation. The world is not tame; only our everyday imaginations are so. It is science itself which has taught us this.

Materialism, taken literally, is the absolutely non-romantic view of nature, the absolute denial of the basis of thought, which is sympathy. If materialism has been felt to be inspiring by some it is because they have not literally meant what they said. When one looks at their definitions of matter, one finds them asserting little more than that the world is largely independent of the human, or any closely similar sort of mind, and this independence is indeed, according to some authorities, e.g., Popper, an assumption of science. But it is also an assertion of most psychicalists.

Another supposed but illusory result of science is determinism. Here too we face a trilemma, a choice among two sorts of monism and a dualism. Freedom, in the sense of relative causal indeterminacy, is non-existent, or all pervasive, or existent but localized. The romantic view plainly is the second or monistically indeterministic concept, according to which all reality has at every moment some range of open possibilities,

98

some spontaneity, or self-determination here and now, and not merely determination by the causal past.

It may be objected that this doctrine sets limits to the scientists' search for causal explanations enabling us to predict the future. But science is not necessarily prediction; it is search for the hidden but indirectly observable intellectual beauty of the world. A strictly deterministic cosmos, with all conclusions foregone, conclusions save for ignorance is not really beautiful. Several excellent recent works on aesthetics suggest a contrary view, pointing out, for instance, that all over the world the preferred musical performers are those who keep doing the not-quite-expected; in other words, who preserve their relative unpredictability. And it is plain for all to see that the zest and beauty of the life of inquiry itself is partly due to the unexpectedness of its results. To foresee scientific discoveries would be to make them beforehand.

Physicists in general now seem happy with their statistical laws. Certain philosophers and a few colleagues may tell them they should not be happy, but they feel that it is a superlatively beautiful world of order-in-disorder which they have discovered. Einstein's inability to sense this beauty rendered his last period, so far as a layman can make out, singularly unfruitful.

The drive for unity goes even deeper than that for prediction. Deterministic systems have always been dualistic, either overtly or covertly. Only indeterminism can be universalized without paradox. It takes order as a matter of degree, a channeling of spontaneity within wide or narrow limits. This channeling can be provided by spontaneity itself. Freedom in each case is thus limited only by the freedom which has been already exercised in past events. Today I am limited by my choices of yesterday, and by past choices of my friends and enemies, and by my cells' or atoms' trifling little choices, and so on. Freedom as social, sympathetic, concerned for the freedom of others, is all we need in order to explain law and order. To carry out this sublime thought we may need to conceive freedom not only as present in many degrees and on many levels but as actualized on a supreme or divine level, by which all lesser freedoms are coordinated sufficiently to keep conflict within bounds. This idea too is grand and beautiful enough to satisfy the spirit of science, even though it transcends empirical tests.

We can now deal with the supposed truism that science has set aside values, "final causes," in its search for the truth. It has indeed set aside man-centered values, flattering to the individual or the species; but only because it has found that the greatest value is reached through man's forgetfulness of himself, and even his species. What pitifully unimaginative values were enshrined in some of the old doctrines of final cause! The sublime universe arranged just for our species, each atom put precisely in its place, devoid of the least suspicion of spontaneity, a world of puppets—or rather, of things even less free than puppets, since, as physics now shows us, real puppets are controlled only in their gross behavior, their invisible micro-constituents being neither pushed nor pulled but rather responding to "stimuli" in a way which we have no reason to think is causally determined, though it is causally conditioned.

But the greatest pettiness and Narcissistic blindness in the old theory was in the notion of cosmic justice, as a bookkeeping system of rewards and punishments, the books being everylastingly balanced, and yet inconsistently tampered with by divine mercy. What is this but a cosmic projection of our human legalistic concepts of rewards and punishments, which, we are learning, work none too well even in their proper sphere? And the whole rests on a confused metaphysical assumption, that there is or even could be an entity called the soul or ego which is always identical, and which, committing an act at one time, can be rewarded or punished at another. Science has found that the notion of individual substance or thing is dispensable, that the concept of events, and their relationships and qualities, suffices everywhere, while that of thing does not even apply, on the ultimate micro-level.

Buddhist tradition anticipated this result in its "no-substance" doctrine. Modern psychology supports the view. The attempt by some psychologists to reintroduce the soul as a single entity is rightly regarded with suspicion, or as retrograde. From Hume to James and Whitehead the best modern philosophers have, in one way or another, sometimes rather against their will, testified to the uselessness of this concept, except as a crude approximation to what can be said more precisely in terms of momentary selves, or experiences as Jamesian or Whiteheadian self-active subjects. If the unit of reality is the event or the momentary self, then rewards and punishments have no ultimate rationale, save the purely pragmatic one of influencing conduct. The self which is rewarded is never, strictly speaking, the one which did the deed. Metaphysically, virtue must be its own reward, but in the present, in the act itself.

The religious commandment to love the neighbor "as oneself" fits the Buddhistic or Jamesian doctrine. For if the future experiences of *this* organism are numerically different subjects from the present experience, there is a kind of altruism or love in taking an interest in "one's own" prospective welfare. But then a similar and broader relation of sympathetic foresight can cause one to act for the future welfare of another organism. Thus love for the other person is love for a numerically different entity only in the same general sense in which "self-love" directed upon the future is also concerned with a numerically different entity. Christian philosophy, by its acceptance of the Greek concept of substance, has been forced, over and over again, to explain that we cannot literally love another "as ourselves," since in the one case there is sheer non-identity of substance, in the other, sheer identity. Buddhists, followed by recent Western metaphysicians (for

101

example, Whitehead), see that the non-identity and the identity are alike relative rather than absolute. In a sense we are one with those we love, and not one with our past selves which we have outgrown, or future selves not yet actualized.

Empirical science is not the whole of knowledge. There are at least two other forms of inquiry. One is pure mathematics. Unlike natural science it does not fit ideas to facts, but ideas to other ideas through logical relations of necessity, compatibility, and the like. Like all knowledge, mathematics has its source in romance, in this case the yearning for the beauty of abstract relationships.

There is at least one more form of knowledge. Like mathematics, it relates ideas to other ideas, and not to actual facts, but it deals with a special kind of idea and a special kind of relationship. The ideas—called metaphysical—are those of such generality that all other ideas presuppose them or their equivalents. They have two other features inseparable from their generality: there are no alternatives, in the sense in which blue is alternative to red or green, or lion to tiger, or hyperbolic to Euclidean geometry; moreover, the ideas cannot fail to be instantiated in existence of some kind. Not that there is any particular fact which the ideas require, but that any possible fact can only be an instance of them. The alternative quality to red is green or some other quality; but no matter what things are, they will exhibit the metaphysical ideas.

I shall give three examples of a metaphysical idea. First, the idea of event or happening. Any idea presupposes this idea. Whitehead, who (considering his profession) must have known something about the matter, argued that mathematical ideas all express possible features of events, and nothing else. More concrete ideas than the mathematical rather obviously presuppose the idea of events. Again, there is no idea alternative to "event" as red is to green, or lion is to tiger. In place of red one might have had green, in place of a lion a tiger. But in place of events? Nothing can take their place. Finally, to suppose that there might simply be no events is to utter words without any clear meaning. It is absurd to look about to see if anything is happening. The looking is already a happening or set of happenings. The reality of events is the presupposition of any hypothesis and any observation.

My second example is the idea of value, no particular value, but value as such. Not, of course, necessarily human value, value to members of our species, but simply value to someone, or some valuer, perhaps to an oyster, or to God. All our ideas presuppose this one; for, if an idea has no value, we cannot be thinking it. Again, there is no alternative: for how could anything take the place of value? In any conceivable situation there must be some value. To think a world is already to love it at least a little, and hence impute value to it.

Our third example is the greatest of ideas, that of God. This idea has been formulated in many ways, not all of which are internally consistent. The most usual formulations involve more or less obvious inconsistencies (euphemistically termed paradoxes); however, a minority of theologians and theistic philosophers seem (though this is controversial) to have found a way to avoid these paradoxes. (In so doing they also escape the famous problem of evil, at least in its most menacing form.)

Since no observation of fact is needed to show that the self-contradictory is untrue, logical analysis seems sufficient to dispose of most forms of theistic philosophy. But the untraditional view of God spoken of has not been shown contradictory. Moreover, analysis seems to detect in this idea all the features of metaphysical ideas in general: (1) it is implied by every idea; (2) it has no alternative) and (3) its realization in existence is logically necessary. Thus we return, after long misunderstanding, to discovery which Anselm made nine centuries ago, that the question of the divine existence is not factual but logical. Recently two authors well trained in contemporary analysis have argued fairly and cogently for this conclusion, one in the interest of atheism, and the other in that of theism.[2] Both hold that the question of God's existence is essentially logical, but the one argues that the logical verdict must be negative, the other that it must be positive. I maintain that both are right: if the idea of God is defined in the usual way, the verdict is negative, for the idea is contradictory; if it is defined more carefully, but somewhat untraditionally, the verdict can be positive. In neither case is there room for factual observation, since the concept, "existing in mere contingent fact," contradicts that of God, whether defined in either of the two ways mentioned. Unfortunately, neither author referred to is (or at the time he wrote was) aware of the possibility or necessity of a non-conventional definition of God. Here as elsewhere prescientific common sense, including theological common sense, turns out to be at best a crude approximation. The romance of inquiry reaches a more sublime revelation than some of the self-styled revelations.

The two ideas of God referred to differ as to the sense in which God is properly conceived as infinite or absolute. That there is such a sense is agreed; but the minority view holds that it is precisely not the sense which the great majority of theologians, as though with an infallible instinct for the lesser truth, have adopted. Such is the human mind, whenever it shields itself from the critical process.

In talking so much about the beauty and romance of science, I may seem to be overlooking the dark aspects of our technologically supported but also technologically threatened and polluted world. I am sadly aware of these aspects. But they concern the way scientific discoveries are used. How they are used depends partly on the extent to which science is

104

viewed as merely preliminary to technology, rather than as an expression of intellectual love for all existence. Materialistic or (much the same thing) dualistic theories of reality take too low a view of science, which is our joy in the hidden beauty of nature, intellectual sympathy for subhuman and (in cosmic aspects) superhuman forms of awareness.

In considering the human condition we need to remember that in our species partial freedom from determination by the past reaches its peak on this planet. All freedom involves risk; technology vastly magnifies both the good and the harm that individual choices may produce. In all species but ours behavior is largely guided by patterns that have been tested through natural selection for myriads of generations; our brain capacity perilously reduces this guidance. In us we have the awesome experiment of a species having to live by thought rather than inherited impulse to a degree vastly beyond the reach of any other species on our planet. Can such a species avoid destroying itself and perhaps most other species with it? This is the question we now face. To discuss it further would be a task for another occasion.

NOTES

1. For further discussion of these aesthetic ideas see my *Creative Synthesis and Philosophic Method* (La Salle, Ill.: Open Court Publishing Co., 1970) Ch. 16; also *Born to Sing: an Interpretation and World Survey of Bird Song* (Bloomington: Indiana University Press, 1973), Ch. 1B,C.

2. I refer to J. N. Findlay and Norman Malcolm. See "Can God's Existence be Disproved?" *Mind,* 57 (1948), 176. Reprinted in Anthony Flew and A. MacIntyre (eds.), New Essays in Philosophical Theology (London: SCM Press, 1955), pp. 47-56. Since writing that essay Findlay, influenced, as he generously says, by my work, has adopted a less negative

attitude toward the Ontological Argument, though he still distinguishes between his "Absolute" and God. For Malcolm's essay see "Anselm's Ontological Arguments," *Philosophical Review*, 69 (1960), 41-62. See also my *The Logic of Perfection and Other Essays in Neoclassical Metaphysics* (Open Court, 1962) pp. 25-26.

Discussions

Discussion Following Lecture by Professor Yang

Hartshorne:

If you notice the last item, the final criterion of what is beautiful in art, literature and music is whether man relates to it; mathematics, whether it relates to the rest of mathematics; natural science, whether nature utilizes it. I would say the ultimate criterion is how all that fits together—how art, science and humanity together relate to nature, and nature to man. But it seems to me that in this account there is a gap between life and what seems to be the non-living in nature. Now there is one physicist named Prigogine who has just published a book called *La Nouvelle Alliance*, in which he's trying to reunite the understanding of life and the understanding of inanimate nature. In life the temporal asymmetry is an ultimate thing; in biology it's the past in the present and not the future in the present. So I have to hope that Prigogine is on the right track.

Lipscomb:

On the relation of beauty or aesthetics to life—and I do

107

think there is a unique relationship—I have a quotation from John Dewey's *Art as Experience* which was actually in my manuscript. "For only when an organism shares in the ordered relations of its environment does it secure the stability essential to living. And when the participation comes after a phase of disruption and conflict, it bears within itself the germs of a consummation akin to the aesthetic." That's John Dewey.

Hartshorne:

There's another phrase of Dewey's I'd like to quote. A very short phrase: "The unbalanced balance of things." You see, that's a perfect expression for dynamic equilibrium which is like what life essentially is.

I would like to see what some of you would do with Prigogine's latest book because to me that comes closer to the way I'm inclined to think than the way both of you have been talking.

Lipscomb:

Well, I think Prigogine emphasizes the deviations which are very far from anything approaching equilibrium and generalizes it in a way which is certainly further than I would, and in that respect I would take a less extreme view than he does. That is at least one comment with respect to Prigogine's arguments. On the other hand, there are aspects of the behavior of matter and other things—and perhaps life itself—which are quite far from the usual steady state that we see, that might be important. That, I think, is one part of the emphasis that he makes.

Dyson:

I would just like to mention as a professional physicist that sometimes the instinct for mathematical beauty can lead you terribly astray. And that has happened even to the great ones.

I think of Heisenberg in particular. It's a great gamble. You have to be a genius to get it right, and even if you're a genius you sometimes get it wrong. Lesser mortals do well, I think, to watch the experiments.

Gover:

I have a question for Professor Lipscomb. You said you could have talked about simplicity instead of symmetry. Can you comment in regard to simplicity as it relates to beauty in science?

Lipscomb:

Well, I mentioned that Crick had said that simplicity was part of the idea when he and Watson formulated the formula of a double helix, but I think what I had in mind particularly was Weinberg's paper of four years ago here.* The title was "Is Nature Simple?" and he was discussing the four fundamental forces in physics and how they have become aspects of the same thing. Symmetries which are broken in the early stages of history of the universe. And there are two views as to whether that's simple or not, but it does seem to be whenever you can reduce a complex physical law to principles of symmetry that it's something that seems simple to me. When you can reduce the law of conservation of energy to symmetry and translation in time and mechanics, it seems to me that's a simpler way to think about it. That's what I had in mind.

Gover:

Can you think of an example where an artist has supplied the missing piece in some understanding of the physical world?

* Steven Weinberg, "Is Nature Simple?" *The Nature of the Physical Universe,* ed. Douglas Huff and Omer Prewett (New York: John Wiley and Sons, 1979), pp. 47-62.

Lipscomb:

Well, I can give an approximation that's at a lesser level, and that is the work of Escher in using symmetry—particularly color symmetry—in formulating his pictures, but he did not make a contribution to color symmetry. Not an original one. That is as close as I can come to that.

Dyson:

I suppose that the most distinguished attempt to do justice was the color theory of Goethe, which turned out to be a rather dismal failure, but he firmly believed that it was possible to reach scientific truth along the route of art. And somehow it didn't work.

Yang:

I would like to comment and in fact discuss this particular question which was just raised from a somewhat different angle. I think we all agree that artists and musicians and writers create a cultural background in which we all live. That that cultural background has something to do with much of what we do is very obvious. But we are specifically addressing the issue of whether that cultural background influences a scientist's aesthetic judgment and thereby influences his work. At least I choose to discuss it from this angle. If we do that, I suspect that the problem is extremely complicated. I would not say it is obvious that the general cultural background does or does not include our judgment about aesthetics or about beauty in our scientific work. I'm not saying that it is a determining thing, because clearly there are many other things, but I'm asking, "Is it of some appreciable influence?" Why do I say this? Because it is clear that when you are at the frontiers of any creative work you are making choices all the time. Many of these choices are made subconsciously. You do not even know that you have made the choice. So we can ask,

110

"What is the structure in the brain that makes you make these choices?" I do not know how deep the concept of aesthetics is in our brain, but I would say it is quite deep. Now, part of that is presumably related to hardware. "Hardware" means how our brain is structured. Part of it is evidently related to software. "Software" is the thing that you build into your brain through childhood, through youth. I think that this is a very complicated subject, and all that I can say is that I don't have any special prejudices. But I would argue that any flippant answer about this specific question is likely to be wrong.

Dyson:
I think it was no accident that quantum mechanics was discovered in Germany. There was an intellectual milieu which was very widespread in the arts and in social thinking of that time which was radical in tone. It involved overthrowing all the established conventions, and I think this did help in putting people in the right frame of mind to make radical innovations in science. To that extent, certainly I think it's true: art influenced science, but not directly. And there were in fact, very strong counter-currents; in fact, there were very strong movements among the intellectuals against precisely these irrational tendencies. It turned out, of course, that in the end to be irrational was the most rational thing, but that couldn't have been foreseen.

General Discussion Following Lecture by Professor Lipscomb

Gover:
My first question has to do with nature's beautiful equations, which were referred to earlier today. There are a great number of beautiful equations, but one of the problems seems to be that only a certain number of these equations are actually relevant. Not all of them describe nature, so you

111

must choose which ones to use. Could Professor Yang elaborate in some way on how one goes about choosing, and what's to be done with all these equations that are left over?

Yang:

It's a very natural question. In some of the cases the beauty of the equations is not realized until later. The meaning of Maxwell's equations, which are most important and most beautiful, was not fully appreciated until recent years because we now see more layers of this beauty than have been expected or suspected before. So in the first place we must remember that the beauty of the existing equations may acquire a different meaning as time goes on. The total number of beautiful mathematical theories or beautiful mathematical equations far exceeds those which have been used in physics. I would roughly say that if you look at the very beautiful mathematical structures that are known today, it's only on the order of a few percent—not more than ten percent—that have found their way into physics. Is it likely that the remaining ninety percent would someday find their ways into physics? I would find it rather not likely because, while some of them may in fact get into physics in future years, the mathematicians are very good at creating even more beautiful things. So there will always be some large areas of mathematics left over which are beautiful, but which are not used in physics. Which area of mathematics is likely to be used? This is extremely difficult to tell. In general, the geometric aspects seem to have a tendency to find their way into physics, but that's not always true. Quantum mechanics is based on the concept of Hilbert spaces—a very beautiful and profound abstract idea in mathematics. It's only partially geometrical. I would say it's not really geometrical, but it is one of the fundamental concepts that physicists have to use. If you had asked people before quantum mechanics, that is 1925, whether physics was

112

likely to use Hilbert space theory in a big way, I think few mathematicians or physicists would have said that that's likely to be the case.

Lipscomb:
I wanted to ask one of the physicists if the theory of singularities and fluctuations isn't the new area that might turn up. I mean, that's where a tremendous amount of mathematical work is completely absent, and this is an area which you need.

Yang:
I certainly agree that theories about singularity and probability theories and partial differential equations also have very beautiful things that are going on in them from the mathematical viewpoint. Some of them, I would think, will find their way into fundamental physics, but it's difficult to guess which ones. A number of years ago—in 1931 to be precise—Dirac wrote a very interesting and extremely ingenious paper. In the introduction of that paper he speculated about the future development of physics and how physics is going to absorb new mathematics or utilize new mathematics. He said that there are two ways in which physics would find new mathematics to use: one was through intimate contact with experiments—and he termed Heisenberg's discovery of quantum mechanics as an example of that. The other way was more through ideas about the beauty of the mathematics, and he termed Schrödinger's discovery of wave mechanics, discovered just a few months later than Heisenberg's great discovery, an example of this. It turned out that the two discoveries are deeply related, but Dirac termed Schrödinger's discovery via the second route as less in contact with reality, with actual experimental results. Then he said, let us look into the future and ask which of these two aspects is going to

become more and more important? He conjectured at that time—a very amazing conjecture too, I would say—that it would increasingly favor the second one because experiments are going to become more and more difficult. Indeed if you look at today's huge apparatuses which cost hundreds of millions of dollars and thousands of man-hours to work on, you would realize how correct he was.

Gover:

Is there beauty in disorder? For example, is there beauty in the kind of thing Cage does and is the disorder an aspect of the beauty?

Schuller:

I was hoping to have a chance to make the point somewhere along the line that in fact there is a kind of beauty in chaos, in complexity. This splits into many little infinitives because the complexity, or the chaos, whichever it may be, in John Cage, being chance determined mostly, being left really to a great complex of happenstances, that's a different kind of chaos or complexity than you get from the kind of virtually totally controlled music of, let's say, a Milton Babbitt. These two composers represent the absolute opposite extremes of a spectrum of musical concepts and techniques. What many people regard as pure chaos in Milton Babbitt's music for me, for example, has an absolutely pure radiant beauty. Now that is, however, a chaos that is organized and controlled; in fact it has often been dubbed "organized chaos."

I want to say one more thing that might add another little subtopic, and that is that I find things in music that are inherently complex and that cannot be reduced to anything simpler. If you reduce it you destroy it. That is something I didn't know when I was very young, but I learned it through experiences in music—empirical experiences.

114

Lipscomb:

I certainly think irreversible thermodynamics—the highly irreversible ones, the kinds of thermodynamics that are very far from equilibrium—would fit into this category on the scientific side to match what Gunther has said. There may be other examples.

Yang:

Instead of trying to answer the question, let me ask Gunther Schuller a question. The complexity you speak about, for it to be accepted as something beautiful, must mean something to a number of people. If it only means something to the composer, or if a painting means something only to the painter, I think eventually it will not sell, the idea will not sell. It must have something which can be transmitted, maybe not by one looking or one hearing, but eventually you will get it. I remember when I first saw Jackson Pollock's paintings, I didn't get anything at all. But after I looked at quite a number of them, I began to see the simplicity of the complex situation. And if that thesis is right, then I see similarities in science. The molecular forces are extremely complicated, but there is something which is simple about this very complicated system, and that's what we would study. A thing which is just complex and has no specific meaning—I would find it hard to regard it as really that interesting, either from the scientific viewpoint or from the artistic viewpoint.

Schuller:

Oh, of course I quite agree. What was considered incredibly dense and complex in Brahms's music in Boston in 1890—so that they had signs "Exit in Case of Brahms"—has long been absorbed into the tradition. It is now still perceived by intelligent musicians or thinkers as both simple and complex.

115

Hartshorne:

I don't think it's that all aesthetic values should be identified with beauty. That stretches the word awfully wide. I think in all intense beauty there is an element of disorder—if you want to use that word. It isn't simple order. Within limits we can enjoy various kinds of unbalance. And then there is a higher order of balance between the balance and the unbalance, the sublime and the ridiculous, and so on, and you want all of them.

Yang:

With the permission of Freeman Dyson, I will tell you a story about him. It may have a bearing on what you were just discussing in the last few minutes. Freeman started his career as an excellent young mathematician, and somehow he decided to switch to physics. He was in Cambridge, England, at the time, and there was a very brilliant young physicist by the name of Harish-Chandra (who, by the way, is now a colleague of Freeman's), who had decided to switch into mathematics. The story goes, one day they saw each other while they were walking across the campus, and Freeman said to Harish-Chandra, "I understand you have just switched to mathematics," and Harish-Chandra said, "Yes."

"Why did you switch?"

"Oh, because physics is too messy."

And then Harish-Chandra said to Freeman, "I understand you have just switched to physics."

"Yes."

"Why did you switch?"

"For the same reason."

Is it true?

Dyson:

More or less.

Gover:

A question from the audience: "Do you agree with Canon's Law, which states that for any given phenomenon the simplest explanation has the highest probability of being correct?"

Lipscomb:

My trouble with that question is that the main problem is to try to think of all the possible solutions that you can. If you just think of the simplest one you could very well be wrong, but it does have a high probability of being right. The real problem is to try to think of all the possibilities because, more often than not, the one you didn't think of is the right one. This is known as the method of Holmes—Sherlock Holmes. I referred to it in one of my papers. We sometimes make all of the possible mistakes before we finally choose only the last possibility. Four times in the complete works of Sherlock Holmes he tells Watson whenever you examine all possibilities except one, then that one must be the right one. But it's very difficult to be sure you've got the last one.

Hartshorne:

That principle of exhausting the possibilities is what I've always tried to do in philosophy, but regarding simplicity, I'm inclined to quote Whitehead, who said "seek simplicity and mistrust it."

Lipscomb:

I find beginning students sometimes believe that the mathematics will tell them something, and I think that's most likely not true. It's the physical intuition that they have to develop, and then they guide that with the mathematics as a *language*. I think it's better to regard the mathematics as a language, rather than as a tool which is going to tell you something. The same thing applies to computers, incidentally.

Gover:

Professor Yang, you have said, roughly, that if one had a choice between the beautiful theory and a discordance with the facts, that the beautiful theory would be chosen, and the facts would have to shift for themselves. If facts do not fit a theory, how does one decide whether to mistrust the facts or doubt the theory?

Yang:

If one has to ask that question, one has already lost the game.

Hartshorne:

I know a charming lady, and when I pointed out a fact that conflicted with what she seemed to believe she said, "Oh, a fact is nothing but an obstacle in my path to a theory."

Yang:

This brings me to a point of some importance which I think we have not discussed. There is a beautiful story told by the great astrophysicist Eddington. He said that there was once a fisherman who was a keen observer of nature. He observed, and after 20 years he suddenly realized that he had discovered a new law of nature. The law of nature was that all fish are longer than four inches, but the reason was, of course, that his net was a four-inch net. Now, as experiments become more and more complicated, we must also realize that experiments are oftentimes very much tied with our theoretical concepts. If our theoretical concepts are somehow wrong, then nobody ever looks into some specific directions. Is it possible that we end up with the fisherman situation? I don't know what the answer is to that.

Lipscomb:

Discoveries are sometimes missed. Frequently they're missed two or three times before they're finally really done right so that everyone believes them. Discoveries in science are often already there in papers that people think are not right.

Yang:

I have a story in this regard: there was a professor at NYU who later went to Johns Hopkins who in the 1930's had discovered something very strange about Beta radioactivity. He and some graduate students worked on this, and together they published many, many papers. All these papers were regarded as totally wrong because, if they were right, parity is not conserved. So nobody believed them, and they were very frustrated by the fact that they were told in no uncertain terms by distinguished theorists that this cannot possibly be right.

Lipscomb:

This is what I had in mind. That's a very important point about discoveries in science.

Discussion Following Lecture by Professor Dyson

Hartshorne:

Maybe as the most ignorant one I will begin. This talk comes much closer to my concerns than the previous talks because I've been worrying about the relations between biology and physics all this time. A long time ago, I think it was J. S. Haldane who said, if biology is ever reduced to physics it will be because physics has become more like biology, not because biology has become more like physics. That's the prejudice—if it's a prejudice—that I have. I think the final measure of understanding is the analogy with our own expe-

119

rience and our own selves. You know, a long time ago Plato said that the animal is the model for thinking—the whole cosmos was a kind of ideal animal according to Plato—and I still cherish that idea that the cosmos is alive and not dead. But there's another aspect of this. A good many people, beginning with Leibniz perhaps, have held that this analogy applies not only to other people, and to the other higher animals, and to the whole universe, but it applies in a simpler and more primitive form all the way down through the hierarchy of animals down to cells, molecules and atoms. And if you take the view that atoms do things which are not the result of any particular stimulus, apparently, this accords with Whitehead's metaphysics—that every really unitary constituent of nature is making what amounts to a decision. Wheeler's view seems to fit that. Everywhere things are done that are not literally prescribed by any law, although they fit certain statistical laws.

Lipscomb:
Well first I want to express my agreement with Professor Dyson on the two diverging views, and if Professor Yang and I have overemphasized the unifying one it's just the way I like to think about things at the beginning. But with respect to biology and physics there's a subject in between—chemistry—and the present trend is to try to understand biological phenomena in terms of chemistry. In fact there are other grand unifying concepts besides the DNA one—the nature of how enzymes work, which is one of the subjects that I've been working on for 20 years now. This is just one example of the explanation of a large body of biological phenomena in terms of chemical principles, and of course underlying the chemical principles are the physical ones. But it's the very diversity of things that makes not just biology and chemistry but also physics so complex eventually.

But I would like to raise another question. Are we able eventually, pursuing these ideas, to understand all the universe? It's by no means clear that when we pursue unifying or diversifying ideas that we will be able to encompass all of the things in a description or in our thoughts about the universe. We're doing it now because it's successful, but ultimately we don't know the answer to that, and that's a nice question to raise. I don't answer it.

Dyson:

Well, there is a famous saying of Thorstein Veblin that the outcome of a successful piece of research is to make two problems grow where one grew before. I think that will be always true, so I'm not a believer in understanding everything.

Schuller:

I was very moved by your address, and inspired by it, and it kept me thinking a lot about parallels between the fields you were talking about and mine in the arts. What I was particularly struck with is that we in music also have the need for both unifers and diversifiers and that we have had them in considerable abundance. We've even been fortunate enough to have quite a few who have been both, which is a neat trick. I would think that a man like Mozart was a unifier, whereas a man like Beethoven was much more in the direction of a diversifier. But then you have fascinating examples of someone like Arnold Schönberg, who was both a unifier and a diversifier. It's very difficult to define even now at some considerable distance in time Schönberg's contribution to music—whether he is an evolutionist or a revolutionist. Well, the fact is he was both.

Gover:

I think perhaps people would be interested to hear in what way you feel that Mozart was a unifier and Beethoven was a diversifier.

Schuller:

It's certainly not a great point, and it's not original with me. Mozart came along at a marvelously propitious time when a great amount of experimentation and innovation had taken place in the previous century and a half or more. He was, in effect, handed the basic ingredients, virtually on a silver platter (this can be said similarly of Haydn I would think), and he was then able to unify all these diverse embryonic elements into a whole—a new whole which then became a new plateau, a new platform for further developments. Beethoven was, of course, the greatest beneficiary of that. So it's in that sense of pulling together an enormous amount of previously excavated, extrapolated, innovative material and seeing the coherence amongst all of that which the others before him could not see. Then somebody comes along and breaks that apart again, so it's this cycle of unifying and diversifying.

Yang:

I would like to ask Professor Dyson whether he has any remarks to make on the following question. The characterization of some creative workers as diversifiers and some as unifiers is very interesting. Would you care to comment on how a person chooses to become a diversifier or a unifier? Does a person choose to become that before the age of 20, at the age of 30, at the age of 40? Is it an interaction, the result of an interaction between him and the subject matter he chooses to study, or is it somewhat more inborn before he reaches maturity—say, at the age of 20?

Dyson:

Yes, I would. I have never given any thought to this question until this moment, but I would say off the cuff that the answer is, it must be inborn. One takes the case of Einstein, who we know from his early writing from the age of 14 was already struggling with the question of ether, trying already to come to grips with the underlying principles of the electric and magnetic fields. He had from the beginning the passion to understand at the very deep level. The passion to understand is, I think, characteristic of unifiers, whereas the typical diversifier is somebody who is in love with the object itself, and I belong to that category. I gazed at the stars as a young boy. That's what science means to me. It's not theories about stars; it's the actual stars that count.

Lipscomb:
I too started in astronomy and gazed at the stars, and I wanted to know how they worked. So I went somewhat the other way. But I have a, I hope, correct story about Einstein in this particular connection, that his father gave him some magnets, and he felt force between the magnets, but they didn't touch each other. He wondered what was happening in the space between. I don't know whether the story is true or not, but it probably is. Every person does this, but he thought about it. That's the difference between a scientist and one who is not. A scientist often thinks about very ordinary things that everyone sees, but in a different way.

Hartshorne:
I'd like to make a personal statement about this. I read Emerson's essays when I was 15 or so, and he interested me in metaphysical problems. I'm known as a metaphysician. I suppose a metaphysician is a unifier if anybody is, but just for the same length of time exactly I've had an interest in ornithology, and I'm actually known as an ornithologist now in a fairly official way. That seems a rather opposite thing. There are thousands of singing birds, for instance, and of course, you're interested in the object itself for sure. Every bird has a distinctive song. But since I was also a unifier I was looking for theories. Birds sing, animals sing, insects sing, frogs sing, and whales sing, and even, I think, howling wolves sing. So I

123

put the question, what do they all have in common? I think I have found—and some specialists agree—some principles that apply to all the singing.

But there's one more thing. Metaphysics as a unifier, I think, is logically in a different class from other unifications. There's a technical sense; it's the sense that Karl Popper talks about, in which metaphysics is not empirical. I agree with Popper on this in a definite technical sense. It's not vulnerable to negative evidence in any conceivable experience, so the question is, what makes sense, not what fits the facts. You're concerned with principles about what would make a fact a fact—any possible fact. And so that's in a different class. So I claim to have dealt consciously with two logically different kinds of knowledge.

Gover:

Professor Dyson's talk has drawn a distinction between unifiers and diversifiers and has invoked a statement of complementarity, which is well known to scientists. How far are you willing to push the idea of complementarity in this particular construction?

Dyson:

Well, not very far. I think I used the word really as a metaphor here, and I don't intend that it should have the full technical meaning that it has in physics. So I would like to take it only as a way of speaking, and I don't want to put too much weight on it.

Discussion Following Lecture by Professor Schuller

Hartshorne:

I do want to express some agreement now with Mr. Schuller's global view. I've been inclined to think of it that way. It

124

seems to me that music is a universal language in a sense in which ordinary language isn't universal at all. You can make something of almost any foreign music if you're sensitive enough, and you don't have to be taught about it. I'd go a little further. I think we can make musical sense out of a good many bird songs; I think they speak to us all over an enormous gulf of radically different types of being.

Yang:

Mr. Schuller raised so many extremely interesting and stimulating and insightful points. I would merely comment on one of them. I understand you to say that you believe there is, in fact, not just one system of music; there are many many possible systems of music. What that means, of course, if we return to the theme of this conference, is that the judgment of what is beautiful in music is not unique. It is quite varied; it is culturally dependent. And if we could imagine another planet with living creatures on it, their system of music may be entirely different from ours. If we accept that, we can ask the same question about science. Are there many systems of physics possible, and is the judgment of what is beautiful in physics unique, or is it dependent on what particular system you are talking about? I would be bold enough to venture an opinion on this. I think in the natural sciences, in physics, in mathematics, in chemistry, and in biology, the system is more unique. Suppose you go to another planet with very intelligent, advanced people on it. Say they know radio communication. They may not write Maxwell's equations in the same way that we do, but that they know Maxwell's equations I would not doubt. The same thing may not be written down in the same fashion; they may not be taught in the schools in the same order. Clearly they would be using a different language to describe everything anyway, but that the judgment of what is beautiful in it I would doubt is not essentially the same. If this thesis is also correct, then I think that we have a very fundamental difference that we could pinpoint on the question of beauty in the different disciplines.

Lipscomb:

May I try to simplify, to make a communication between these two sides? It is very difficult to imagine any other kind of life not perceiving symmetry as we discussed it, or not having the same value for pi in the arithmetic system. I mean, one state legislature tried to pass a law that pi should be 3. That doesn't work. In that sense the laws would be much the same. Now it is very hard for me to imagine a kind of music in which at least some of it has not a relation to ours. It would be hard for me to imagine that there would be a kind of music on some other planet where the overtone series is totally missing. Now I've given an example which I believe would have to be universal. Do I produce a connection?

Schuller:

I would think that another system of acoustical phenomena might be out there in the rest of the cosmos. Might be and probably is. One stumbles on those words, but I cannot imagine that our particular overtone series which does very much govern our art is unique.

Are we then saying that the physics of this planet are surely those that govern the entire cosmos?

Yang:

Physics, yes. If that yes answer is not accepted, all the physicists would lose their jobs immediately.

Schuller:

I wish we had something as clearcut in music.

Yang:

May I say further that I would make an added comment on the point that we are just discussing, and here I am completely out of my depth. All that I know about neurophysiology is what I read in the *Scientific American,* but I do have friends that I talk to, and I learn that it's a most exciting field. Its results in the next 20, 30, 40 years may go a long way in clarifying some of the points that we are discussing and that have also been raised by Mr. Schuller at the end of his talk. In particular there are two conspicuous things about the structure of the brain that they are now finding and concentrating on. One is the fact that our senses send signals to the brain in a very complicated fashion through many layers of integration, so that the first response of the retina or the ear to the external stimulations is not the one that goes into the brain. Information is integrated and processed through several layers, and this processing is deeply related, clearly, to evolution and to how the brain is structured. That's one point, namely, that the complexity of the information that reaches the brain goes through hardware which our evolution evolved. The second point is that in the brain eventually there is some mechanism in the reticular center that makes choices. These choices are remarkable because we have signals from all channels reaching our brain, and yet our brain has the miraculous ability to concentrate on one specific thing and turn everything else off. In other words, there's a priority judgment mechanism in the brain, and that mechanism clearly has something to do with all our thinking processes, with all our judgment of what's worthwhile and what's not worthwhile, what's desirable and what is not desirable, and what is beautiful and what is not. It would seem to me that the structure of the brain is very much related to our appreciation of what is beautiful in the arts, in music and in literature. But I would come back to the original theme that nevertheless there are some things which are more absolute in the physical system, in the scientific system. They relate to things which are outside of our brain. They relate to physical phenomena.

Hartshorne:
Isn't that essentially the same as saying that scientific ideas are subject to falsification through observation?

General Discussion Following Lecture by Professor Hartshorne

Dyson:
I'd like you to spell out the two definitions of God mentioned in your paper, one for which the ontological argument is true, and the other for which it is false.

Hartshorne:
What I call classical theism was formulated in two words by Aristotle as the "unmoved mover"; it's the immutable cause of all becoming and all change. It influences everything, but nothing influences it. So if you pray to that god, you can't possibly have any influence on it. By definition you can't have any influence, yet there were lots of prayers addressed to it. The result is rather paradoxical. And worse than that, that god was supposed to know the world. Now Aristotle knew better. He knew that that kind of god couldn't know the world because to know something is to be influenced by it. So Aristotle denied that god knows the contingent details of the world. But in the Middle Ages they felt religious obligations to a certain god who knows all about us. So they had Aristotle without Aristotle's limited consistency. Spinoza saw that problem, and therefore he denied the contingency of the world so god could know everything without being contingent in any respect.

So that's a long dialectic covering many centuries. But soon after Spinoza the Socinians decided that they'd solve the problem in a different way. They asserted the contingency of the world and asserted that god knows the world; they said therefore there's something contingent in god, namely the world as he knows it. And since we are free, we have power to determine, to some extent, what God is going to know about the world. We influence God. So the Socinians were the first process theologians. And Fechner in Germany had a similar idea. Peirce came along and hinted at such an idea. Bergson seems to hint at it without ever making it clear, and Whitehead tried to develop it systematically. So that's a tradition from 1600 to the early twentieth century down to the way that I think about it. Those are the two basic ways of thinking about God, and the second way gets out of some of these paradoxes. God knows the contingent world because he has contingent contents in his awareness, and God knows the changing world, but God changes as the new creatures in the world change. There are new items in God's knowledge, so there's novelty for God as well as for the world. This is a basically new idea.

Schuller:

Well, I had one sort of small comment or point. I was, of course, fascinated by your talking about the relationship between predictability and unpredictability, and other such opposites. I of course said earlier today that I find the greatest works of art contain both in some creative balance. But then you quoted Kurt Sachs with a quotation which I couldn't quite agree with. I think it read, aesthetic order is the vast realm between the fatal extremes of mechanism and chaos. I feel that this is, in itself, a kind of value judgment on the part of Kurt Sachs, and very conditioned by his own background and viewpoints on what constitutes art and what constitutes these extremes of what he called mechanism. I think what has happened since he said that—at least what we've come to realize—is that these borders, these "extremes" of mechanism

129

and chaos, are not stationary. It's an expanding universe, and that to me somewhat sabotages the possibility of making such a statement. Do you follow what I'm saying?

Hartshorne:

Yes. I think this is somewhat of a semantic difficulty. He wanted to exclude the idea that absolute order would be aesthetic and that sheer absence of order would be aesthetic, but those extremes are so extreme that it's pretty hard to get them actually. I think he's allowing for many gradations of good and bad in between those absolute extremes.

Schuller:

But those extremes right at the edge of the visibility or audibility are not necessarily to be given a different value judgment than that which lies between those extremes. In other words, he says aesthetic order is the vast realm between these extremes, implying that the edges of those extremes do not fall into a concept of order, and I'm sure he meant by that beauty. So I'm saying I think we've got to think about including the outer edges, extremes, of such concepts as also being capable of order and beauty—an order and beauty which we may not at first see or conceive.

Hartshorne:

I still think it's probably partly just a matter of words. For me the important thing about this is that if you say something is not beautiful you might mean either of two opposite things: not beautiful because it's chaotic, or not beautiful because it's a mechanical order, which would be very different.

Lipscomb:

I have a question for Gunther Schuller, and I hope that's not out of order. He said there is good music and bad music, and I would like to ask him how you tell which is which?

Schuller:

Well, I think I made it abundantly clear that we do have trouble answering questions like that. We've been struggling with them for 4,000 years, and we do not have answers that we can consider in any sense definitive or absolute. In a very practical sense there is a correlation between good and bad and certain other aspects of creativity which have to do with function, with purpose, with such elements as talent, and perception, and sensitivity, that go into that creation. We tend to evaluate anything that comes new into that creative lineage by those previously established criteria, which, interestingly enough, have been fairly constant for many centuries now.

Lipscomb:

That's good. It's the same in painting. The problem arises when these things that do survive are isolated from their previous contact with humanity and put in museums or something of that sort. Or in science, if the discovery is taken away from its surrounding context, then you lose some of the aesthetic judgment.

Gover:

Would Dr. Lipscomb and Dr. Yang and Dr. Dyson agree with Dr. Hartshorne that up to the present time American scientists have mainly contributed to experimental facts instead of theories?

Hartshorne:

I said "until recently"; I don't say this is true in recent decades. Nevertheless, I would like to hear what they say.

Yang:

Well, I was in fact very struck by that statement. I think that we have to go back several decades before that statement could be considered to be true. The truth is, of course, that

131

the United States in the nineteenth century and the beginning of the twentieth century built up a very strong industrial base. From the industrial viewpoint this country was very advanced and had an infrastructure that was very strong. But because of tradition, pure science was not very much developed, was not very much emphasized in this country. However, that was changed overnight by the war. Once the United States government and the people perceived that pure science is important for society, the United States concentrated its resources. And then there are various other aids in this process including Mr. Hitler and also the Russian Sputnik. All these created a situation in American science that made that statement that you made not quite right, certainly say, after 1945.

Gover:

Since we're on this subject, I have another question. According to a question received from the audience, Werner Heisenberg said the great scientific contribution in theoretical physics that has come from Japan since the last war may be an indication of a relationship between philosophical ideas in the tradition of the Far East and the philosophical substance of quantum theory. Do you have comments about that question?

Yang:

Well, I have never seen that quotation, but that does not mean that I claim to have read everything that Heisenberg wrote very carefully. I doubt very strongly that it's an actual quotation because I cannot imagine that in any period Heisenberg would say precisely those words, because much of it is not true. It is, of course, correct that Japan made tremendous contributions to elementary particle theory through the 1935 paper of Yukawa and the papers of Tomonaga, and Yukawa's and Tomonaga's students, throughout the '30s and '40s. It's also especially significant that these were done at a time when Japan was, in many senses, quite isolated from the rest of the industrialized world. As to the possible implications of the Oriental training in the background of Yukawa and Tomonaga on their work, this is a subject about which there have been a number of discussions. In particular, Yukawa attributes very much of the Oriental influence to his work. I think it's a debatable question, in my opinion, despite the statement of Yukawa's.

Does what I just said contradict what I was saying yesterday that perhaps our judgments of beauty and style are very much dependent on the culture? I would like to maintain that there's no disagreement because the kind of choices that one makes in choosing problems in elementary particle physics—choosing the methods of attack, regarding what is important, what is unimportant, what is beautiful, what is not beautiful—in my opinion have very little to do with the difference between Oriental philosophy and Occidental philosophy. So I would maintain that, in this particular case, I would disagree with Yukawa.

Gover:

Dr. Schuller, let me ask you a question. Someone from the audience asks, "On the one hand, you say that we are in a period like the end of the eighteenth century—the exhaustion of experimentation before a period of creativity. But did not this creativity come about by choosing some forms, developing them and rejecting others? Does not creativity require a selection? If so, creativity stands opposed to the musical pluralism you propose."

Schuller:

It's very difficult to be a composer today for lots of reasons, including the difficulty of ever making the basic choice of what kind of composer, and what kind of technique, and what kind of media, and so on, you might choose to work with. That was a kind of question which, in that magnitude, Mozart could not have possibly asked himself. But there is a confusion in the question. Perhaps I didn't make myself clear enough. I don't see any contradiction because when I spoke about pluralism in music, I wasn't saying that everybody, all of us, have to use all of those things that are about us. I was speaking more in a sense of at least appreciating the existence of these things and appreciating them in their content. I mean, whether I use Japanese court music in my music is, of course, a very personal question.

Gover:

I have a question about language. This person is again quoting Heisenberg, but the question begins as follows: "Is not the language used by modern scientists similar to that of philosophers?" And then he or she goes on to quote Heisenberg. "The problems of language here are really serious. We wish to speak in some way about the structure of atoms, but we cannot speak about atoms in ordinary language." Could there be some comment about the use of language in describing things which are very difficult to conceptualize in terms of concrete things? Professor Dyson?

Dyson:

Yes, well our language is mathematics, and that of course is wonderfully appropriate for reasons we don't understand. It actually does describe what we see going on in the experiments in a beautifully economical fashion, and it's not the language of the philosophers. Philosophers have a very much more difficult problem trying to translate this into words. That's something physicists usually don't even try to do, and there's no reason we should.

Lipscomb:

My language is my visualization of what molecules are doing either in their structure, their transformations, or their reactions, and I translate that either into chemical language or into mathematics, but not into English. It's surprising how little one uses English in the actual working out of science. Most people who are not scientists believe that they think in terms of language. I'm not quite sure that they do. I know I don't. I later put it in English, but it's the third stage of the process.

Gover:

I have a related question: "Gentlemen, would you comment on the use of metaphor—for example, field theory in science—and on analogical thought as a link between science and the arts?"

Yang:

I suspect that the person who asked this question is struck by the very strange and sometimes fanciful names that scientists give to a number of the things that we talk about these days. I think only in very rare cases are they metaphorical. I think they are usually somehow "poured out of the hat." So I would not necessarily attach a deep significance to them. It is true that oftentimes in retrospect one finds that perhaps a better name would have been much more appropriate, but usually it's too late.

Gover:

So "quarks" and "color" and "strangeness" are fanciful terms then, picked out to amuse as much as to inform.

Lipscomb:

Well, I'm not sure. I think it's nice to pick out a name that people can remember, that doesn't have some other meaning already in science, so there may be some way of keeping it from too much confusion if you pick a fanciful name.

135

Yang:

I would answer your question positively. I think in most of the cases the person is trying to be either humorous or cute, if you want to put it that way, rather than to convey some specific information. That's at least the case in most of the terms that I find in the new physics.

Dyson:

I would just like to say that, to my mind, the greatest similarity between science and art is in the craftsmanship of it—which has nothing to do with words—and that also has a great deal to do with the question of how you tell good from bad.

Final Discussion

Gover:

If they would, I would like our guests to reflect on what they think about the aesthetic dimensions of science, what they think the conference has meant to them, make any additional comments that they haven't yet made, or pose a question for one of their fellows here.

Yang:

I'm a theoretical physicist. And indeed I and my colleagues in theoretical physics talk about symmetry all the time, and we sometimes irritate our experimental colleagues by this. When they are really irritated, they tell us a story to take us down a few pegs in our great excitement. I think you might enjoy the story. A man carried a large bundle of laundry with him, desperately wishing to have it done. He went around the town, but couldn't find a laundry anywhere. Frustrated, he was delighted to find at last near a small alley a shop with a sign above it which read "Laundry Done Here." He entered the shop and dumped the bag on the counter. The man behind the counter said, "What is this?"

He said, "Oh, I want the laundry to be done here."

The clerk responded, "We do no laundry here."
"But you have a sign saying, 'Laundry done here.' "
"Oh that. We only make signs."

Lipscomb:
I gave my lecture as a trial to a class at Harvard and they liked it very much. But the problem is that to really present a particular case you have to give all the background of the time in which the particular crisis arose so that the resulting solution can be seen in context. I was making a case earlier that this is no less true for any work of art. It shouldn't become isolated from its generation, because it's in the creative process that the aesthetic impulses are most easily seen.

Hartshorne:
I do have one little point. I was terribly embarrassed in hearing the two talks about symmetry, because there's a puzzle in it which really baffles me. I think it's partly my ignorance and partly probably stupidity, but you see the basic laws of physics are symmetrical, and symmetry is a terrific principle, it seems almost an ultimate principle, apart from agreement with experience. But, on the other hand, if you look at formal logic, the basic principles there are asymmetrical, like implication. You see, the symmetrical case is equivalence; well, equivalence just gives you tautologies. So if symmetry, or symmetrical relations, are the basic principles, it seems that reality is a giant tautology. In biology the basic principles are not symmetrical. Growth is a one-way thing. In psychology, memory and perception both give us the past; we remember back into the past and we perceive back into the past.

Dyson:
Well, I was going to tell a little story—a true story—which I remembered when we were discussing the question of whether the language of science is universal. This is of course

a big difference between science and music: there are very different musics in different parts of the world, whereas there's only one science, and all of scientists more or less understand each other. It's a very interesting philosophical question. Suppose there were some celestial colleagues up there in the sky; suppose that some alien species were also looking at the universe and interpreting it in some sort of fashion. Would their science be the same as our science? Would we be able to communicate with them? This is a question which many people have speculated about, and, of course, we have very little evidence. However, I have had one experience which I think is relevant. I know very well a family who has an autistic child; an autistic child is the closest that I've ever come to an alien intelligence. It is something very, very remarkable. You have a child who is in some respects extraordinarily intelligent, but who looks at the world in a totally different fashion from us. (She's now grown up, by the way, and has become more or less a well-functioning human being.) When she was twelve years old, she spoke very little, and, as autistic children very often do, she lacked the sense of her own identity. She could not tell the difference between "I" and "you." She would use the word "I" and the word "you" quite indiscriminately, not knowing the difference. Well, a letter came for this child from one of her friends who was also autistic, also twelve years old, who lived nearby. So she was very happy, and she opened the letter eagerly to find quite a long letter which had nothing but numbers in it. I looked at it and I could tell right away that these were all the prime numbers from 2 all the way up to 1000. So this child who received the letter glanced through it very fast, and then quickly she took a pencil and crossed out the number 703. I was rather surprised at that and said, "But isn't 703 prime?" She didn't actually say anything, but she quickly wrote down "19 × 36." So her prime numbers are the same as ours, and that leads me to believe that perhaps there is something universal in mathematics, and perhaps also in science.

Yang:

I would like to comment on Mr. Schuller's remarks about the emotional aspects of a composer's work. It is true that there is less emotional involvement of the type that you refer to among the scientists, but it is not entirely absent. When one is close, when one feels that one is close to a very important secret of nature, there is a feeling of fear. I think that feeling is quite understandable because it is as if you feel that you are taking a look where you should not—that this is something which is not naturally revealed to mankind, and here you are now taking a peek. It's a deeply religious feeling.

Schuller:

Yes, I agree. All creative activity has, must have in it, as part of that process, these kinds of emotional aspects.

Hartshorne:

What you just said reminds me of a phrase of Whitehead's: "the imaginative muddle that precedes creation."

Dyson:

Yes, I think that the analogies between science and art are very good as long as you are talking about the creation and the performance. The creation certainly is very analogous. The aesthetic pleasure of the craftsmanship of performance which you have been talking about also is very strong in science. And if one is handling mathematical tools with some sophistication it is a very nonverbal and a very, very pleasurable experience just to know how to handle the tools well. It's a great joy. What we don't have is the listener.

Gover:

I'd like to interject a question at this point. Professor Hartshorne is a philosopher, but he's also an expert on birds and bird song. A member of the audience would like to hear

139

Dr. Hartshorne speak more about bird songs, and how musical elements are perceived across species.

Hartshorne:
   A number of people trained in music, in some cases pretty highly trained, have studied bird song and found it interesting musically. It is, I think, pretty well agreed that the basic, elementary musical devices are all used in bird song: melody, harmony to some slight extent. (Birds are physically equipped to make two notes simultaneously, and they sometimes do.) With the wood thrush, for example, you get a beautiful major chord, and while that won't all be simultaneous, two of the notes may be simultaneous. You also get a minor chord, you get some very nice tonal relations, and you get various rhythms. Songs can have a practically hypnotic rhythm. If that isn't rhythm, I don't know what is. Well, you can get interval inversion, you can get theme and variation—lots of that. Theme and variation is the natural way to describe lots of bird songs. You can get even key change. Some European writer says that no bird does that; I'm not sure that any European bird does, but there's a tropical American bird that does, and there's a bird in Australia that does. There isn't anything simple that some bird can't do.
   Nobody has demonstrated that a bird has a definite pattern that lasts more than ten seconds, and in the overwhelming majority it's under five seconds. That's one limitation. So if they go on and on, that's not a fixed pattern; that's a kind of rambling series, somewhat miscellaneous. On the other hand a whale with a huge brain, bigger than ours, can have a song that takes five or eight minutes to unwind, and that's a pattern. The whale knows that whole pattern. Now it's not a mechanically fixed thing. It's a sequence of themes or phrases, and each phrase or note is repeated a number of times. The number of repeats is indefinite, so the length of the whole thing at the time is indefinite, but the order of themes is apparently fixed. They learn a new song every sea-

140

son, and the song spreads rapidly through the whole herd showing they're highly imitative. And all the best song birds imitate at least to some extent. That means they're interested in what they hear. The imitation shows it, and that's not the only proof that they're interested in what they hear. I wrote a whole book just to argue on the positive side of the question, "Do birds have any musical feeling?"

Gover:

I have a question which is addressed to Professor Dyson. It says, "Why did you expect argument from your colleagues with respect to your lecture?"

Dyson:

Well, it is of course a question of taste, what one finds beautiful in science. One can't easily argue about taste, but clearly my tastes do differ from those of my friends here. They were emphasizing very heavily the beauty of ideas, especially the beauty of unifying concepts that would bring simplification in the world that they are trying to understand, whereas I'm talking much more about the process of discovery, of the beauty and the variety of natural objects that we discover, the beauty of the individual—the beauty for example, of an individual's style rather than the beauty of the theory as to how it moves. The emphasis that I put was deliberately aimed to be in contrast to the emphasis of the others.

Lipscomb:

I at least chose symmetry as a unifying principle because it was easy for me. But I don't disagree at all with Freeman on this point.

Yang:

Could I make a comment? In many respects I have often thought that the way that ones does scientific research is very much like working a jigsaw puzzle. Let me explain what I mean. Of course all examples like this are bound to be imperfect, but there are some striking features which make me feel that the example is an interesting one. When you do a jigsaw puzzle you look at various possible linkages, and somehow one particular form sticks in your mind. All of a sudden you find something which more or less fits what you have in mind which you couldn't find. You say, "A few minutes ago I saw that. Where was it?" You look for it and put it together. Oftentimes it doesn't fit, but once in a while it does, and you're greatly overjoyed. So this aspect of attacking a problem—holding off because you can't make further progress, looking at some other problem, then suddenly having the ideas fall in place—this constant searching for new associations, subconsciously or consciously, is one important element of scientific research. You don't constantly attack one problem. If you have a lot of small linkages, you try to make them fit, and then once in a while you find one piece which can put five pieces together. That joy is just indescribable.

Dyson:

I'd like to speak in a different analogy, but I agree very much with Frank. I have done a good bit of pure mathematics, and when you are trying to construct a mathematical proof it is like building a bridge starting from both ends. You make a great leap and you somehow know, somehow subconsciously you know, that the architecture is going to fit. There has to be this imaginative leap before you can even start.

Schuller:

Absolute similarities, just absolutely no difference. The jigsaw puzzle, the bridge—I've gone through that so many times. It starts with a flight of imagination, yes, a jumping into some unknown territory or area.

Lipscomb:

I was also going to take the bridge analogy and the puzzle

142

analogy and carry them to the aesthetic response that we're studying. You work on the puzzle or the bridge for a long time and you cannot solve either of them. You leave the problem. You may leave it for weeks or months; you may even come back to it. You may leave it again. And all this time, your mind continues to work on it. Then sometimes it happens that the solution occurs to you at a very odd moment that has nothing to do with what you're doing. You're not working on the problem; you're doing something else. Poincaré is one who has described this. Everyone who has done original work has felt it, and that's the aesthetic response. Finally I would say that when I have a problem that has in it an inherent contradiction, or something that is inconsistent, and I really cannot solve it, I have a habit of working on it until about the time I go to bed. I schedule it for that time. If I can't go to sleep—and I usually can't—then sometimes I find that I wake up with a solution although I don't know how it happened. My mind sorted it out somehow. Although we don't understand the aesthetic response we can actually plan our lives in such a way as to maximize it.

Hartshorne:

I was going to refer to Poincaré too. If I'm not mistaken he's the one who, when he was asked how he made a certain discovery, replied, "By always thinking about it." Now this is what led me to the maxim that to be creative you must concentrate. It's as simple as that. I'm convinced that our college systems rather discourage concentration and therefore discourage creativity. You have fifty minutes of a subject, and then you go off and do something quite different; maybe a week later, or at least a couple of days later, you come back for another fifty minutes. The most creative period I had since I was fairly young was in Japan when I gave a seminar three hours every morning with a nice time out to drink tea in the middle, five days a week, two weeks in a row, and the students were not doing anything else. It was between college

143

terms, so we were all sticking to the subject of that seminar and not thinking of much else for those two weeks. I got the biggest rush of ideas that I'd had in 30 years. What came out of that is what I'm probably best known for in technical philosophy. I think our curricular system is rather unfavorable to that.

Yang:

I found Mr. Schuller's description of the creative process of a composer most interesting. There are great similarities in the creative process for a scientist, but I think there are a number of differences, so I would like to ask you a few questions. Suppose you are struggling with a problem. Is it possible for you to communicate with a colleague of yours or a student of yours what you are struggling with?

Schuller:

I certainly can vouch for the fact that it is either never done, or hardly ever done. You're the first person that ever asked that question; in my entire experience I'm not aware of that ever having happened except in some very general way over a coffee table. In the specific sense in which you collaborate, no, and I think there's a reason for that. Our problems are so private, and so ephemeral, and so intangible before they're put down, that even to articulate them to someone else so that he could be helpful seems to be quite impossible.

Yang:

I suspected so. That's why I asked you. In the case of our work we can always explain the end result, the other side of the bridge, as Freeman was saying, so that we can always communicate with our friends and they can also try to build the same bridge. But suppose I didn't discuss it with anybody, and I did build a bridge. I mean the bridge is an extraordinarily out-of-the-way kind of arrangement. Somebody might ask you how you did it. That is where you cannot tell him. You don't know yourself how it came about. It's just that you struggled for a long time and somehow something came, some inspiration came.

The second question I would like to ask you is, well, there are two aspects of our world in theoretical physics and in mathematics; I suspect this is true also of the experimental sciences. You have to have an idea, an inspiration, and then you have to work it out. These are two distinct stages of the work. For example, in our case since we don't have equipment we have to have a pad and pencil and try to derive one formula from another one and see whether the whole thing works. So roughly, to put it very simply, you somehow think of a flight of imagination that would give you an idea, and then you'd sit down and work it out. The question I have is (a) is that also the type of thing that happens in your creativity? And (b) is it true also in your case, as is true with many theoretical physicists and mathematicians, that if you just sit down there and stare at the pad the inspiration doesn't come? Once a newspaper man asked me what is the best time for you to get inspirations. I said when I brush my teeth.

Schuller:

Yes, I think those two things are very similar and very parallel. I think flights of imagination or inspirations of the kind that you describe come to us at all times, unpredictably. They are very much influenced by the environment, context, ambience around us, which is, of course, is everchanging. And because it is so changeable, there's no way of predicting when something is going to hit or from what direction it's coming. What you call the working out period may go in minutes; it may go in hours. It's all very variable, and each composer works differently.